The Memory Jogger™ II

A Pocket Guide of Tools for Continuous Improvement & Effective Planning

First Edition

Michael Brassard & Diane Ritter

GOAL/QPC

This handbook is designed to help you and every person in your organization *to improve daily the procedures, systems, quality, cost, and yields related to your job*. This continuous improvement process is the focus of this book—sharing the philosophy and tools that are fundamental to this effort.

The Memory Jogger™ was first written, edited, and produced in 1985 by Michael Brassard and Diane Ritter of GOAL/QPC. They have, in the name of continuous improvement, worked together again in substantially revising the original tools and writing new material to create a more readable, user-friendly, and comprehensive reference pocket manual. *The Memory Jogger*™ *II* now includes the Seven Quality Control Tools, the Seven Management and Planning Tools, and a problem-solving case study example, which are presented in a format that (we hope) will allow you to find relevant information so easily, you'll *always* find space for it in your pocket or pocketbook.

Edited by Francine Oddo
Cover design by Lori Champney
Layout production by Michele Kierstead
Graphics created by Deborah Crovo

GOAL/QPC

2 Manor Parkway, Salem, NH 03079-2841

Toll free: 1-800-643-4316 or 603-890-8800
Fax: 603-870-9122
E-mail: service@goal.com
Web site: http://www.goalqpc.com

GOAL/QPC is a nonprofit organization that helps companies continually improve their Quality, Productivity, and Competitiveness. (Hence, the "QPC" in our name.)

Printed in the United States of America
First Edition
20 19 18 17 16 15 14

ISBN 1-879364-44-1

Acknowledgments

Our sincerest appreciation and thanks go to our dedicated, knowledgeable, and tireless friends who have worked so hard to make this little book a BIG success.

Production & Administrative Staff: Lee Alphen, Steve Boudreau, Paul Brassard, Lori Champney, Deborah Crovo, Brian Hettrick, Michele Kierstead, Fran Kudzia, Larry LeFebre, Francine Oddo, Dorie Overhoff, & Debra Vartanian.

Concept & Content Reviewers: Del Bassett, Casey Collett, Ann McManus, Bill Montgomery, Bob Porter, & Paul Ritter.

Customers & Collaborators: To the customer focus groups for their insights and suggestions.

Companies Who Contributed Information for Illustrations:
(listed in order of the company name)
Bryce Colburne, *AT&T Technologies*
Robert Brodeur, *Bell Canada*
Karen Tate, *BGP*
Weston F. Milliken, *CUE Consulting*
Paula McIntosh, *Georgia State Dept. of Human Resources, Div. of Rehabilitation Services*
Donald Botto, *Goodyear Tire & Rubber Company*
Tony Borgen & Paul Hearn, *Hamilton Standard*
Walter Tomschin, *Hewlett Packard*
Dr. Verel R. Salmon, *Millcreek Township School District, Millcreek Township, PA*
Jack Waddell, *Novacor Chemicals*
Kirk Kochenberger, *Parkview Episcopal Medical Center*
Marie Sinioris, *Rush-Presbyterian-St. Luke's Medical Center*
Tim Barnes, *SmithKline Beecham*
Dennis Crowley, *St. John Fisher College*
Buzz Stapczynski, *Town of Andover, MA*
Kemper Watkins, MSgt., *U.S. Air Force, Air Combat Command*
Capt. T. C. Splitgerber, *U.S. Navy, Naval Dental Center, San Diego*

How to Use The Memory Jogger™ II

The Memory Jogger™ II is designed for you to use as a convenient and quick reference guide on the job. Put your finger on any tool within seconds!

Know the tool you need? Find it by using the:

Table of Contents. Tools, techniques, the case study—it's in alphabetical order.

Solid tab. This tab runs off the edge of the first page of each new section.

Not sure what tool you need? Get an idea by using the:

Tool Selector Chart. This chart organizes the tools by typical improvement situations, such as working with numbers, with ideas, or in teams.

What do the different positions of runners mean?

Getting Ready—An important first step is to select the right tool for the situation. When you see the "getting ready" position of the runner, expect a brief description of the tool's purpose and a list of benefits in using the tool.

Cruising—When you see this runner, expect to find construction guidelines and interpretation tips. This is the action phase that provides you with step-by-step instructions and helpful formulas.

Finishing the Course—When you see this runner, expect to see the tool in its final form. There are examples from ***manufacturing, administration/ service,*** and ***daily life*** to illustrate the widespread applications of each tool.

Contents

Tool Selector Chart

This chart organizes the tools by typical improvement situations, such as working with numbers, with ideas, or in teams.

Working with Ideas	Page #	Generating/Grouping	Deciding	Implementing
AND	3			●
Affinity	12	●		
Brainstorming	19	●		
C & E/Fishbone	23	●	●	
Flowchart	56	●	●	●
Force Field	63	●	●	
Gantt	9			●
ID	76	●	●	
Matrix	85			●
NGT/Multivoting	91		●	
Prioritization	105		●	
PDPC	160			●
Radar	137		●	
Tree	156	●		●

Working with Numbers

	Page #	Counting	Measures
Check Sheet	31	●	
Control Charts	36	●	●
Data Points	52	●	●
Histogram	66		●
Pareto	95	●	
Process Capability	132	●	●
Run	141	●	●
Scatter	145	●	●

Working in Teams

	Page #	Improvement Roadmap	Team Roadmap
Storyboard Case Study	115	●	
Starting Teams	150		●
Maintaining Teams	151		●
Ending Teams/Projects	153		●
Effective Meetings	154		●

Thanks for the love, patience, and support
Jaye, Nina & Paige
&
Paul, Christian, Karin & Lauren

We owe you one!

Introduction

In classrooms to board rooms, on manufacturing floors and in medical centers, organizations around the world are using *continuous quality improvement* (CQI) as their strategy to bring about dramatic changes in their operations. Their purpose is to stay competitive in a world of instant communication and technological advancement.

These organizations need to meet or exceed customer expectations while maintaining a cost-competitive position. *Continuous quality improvement (CQI)*, a systematic, organization-wide approach for continually improving all processes that deliver quality products and services, is the strategy many organizations are adopting to meet today's challenges and to prepare for those down the road.

In pursuing CQI, stick to these four basic principles:

1. Develop a strong customer focus
2. Continually improve all processes
3. Involve employees
4. Mobilize both data *and* team knowledge to improve decision making

1. Develop a Strong Customer Focus

Total customer focus includes the needs of both external and internal customers. External customers are the end users, internal customers are your coworkers and other departments in the organization.

2. Continually Improve All Processes

- *Identify them.* A process is a sequence of repeatable steps that lead to some desired end or output: a typed document, a printed circuit board, a "home-cooked" meal, arrival at work, and so on.

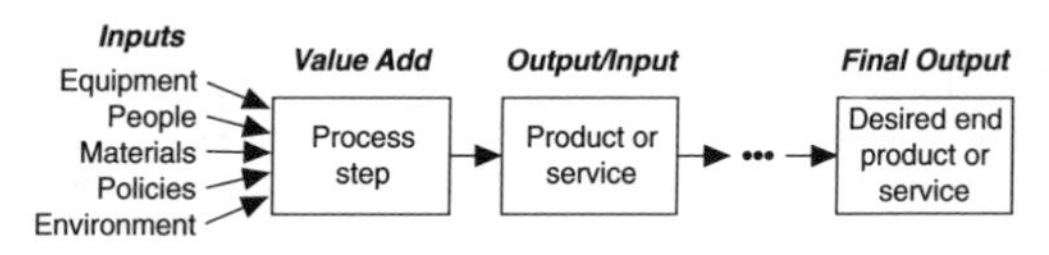

- *Improve them.* Use the **Plan, Do, Check, Act (PDCA) Cycle: PLAN** what you want to accomplish over a period of time and what you might do, or need to do to get there. **DO** what you planned on doing. Start on a small scale! **CHECK** the results of what you did to see if the objective was achieved. **ACT** on the information. If you were successful, standardize the plan, otherwise continue in the cycle to plan for further improvement.

3. **Involve Employees**

   Encourage teams—train them—support them—use their work—celebrate their accomplishments!

4. **Mobilize Both Data and Team Knowledge to Improve Decision Making**

   Use the tools to get the most out of your data and the knowledge of your team members.

   - Graphically display number and word data; team members can easily uncover patterns within the data, and immediately focus on the most important targets for improvement.
   - Develop team consensus on the root cause(s) of a problem and on the plan for improvement.
   - Provide a safe and efficient outlet for ideas at all levels.

Use this book and the tools in it to focus on, improve, involve employees in, and direct your path toward continuous quality improvement.

Activity Network Diagram (AND)

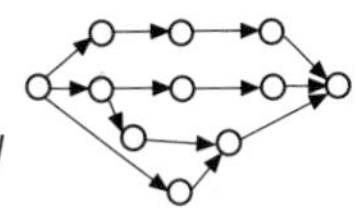

Scheduling sequential & simultaneous tasks

Why use it?

To allow a team to find both the most efficient path and realistic schedule for the completion of any project by graphically showing total completion time, the necessary sequence of tasks, those tasks that can be done simultaneously, and the critical tasks to monitor.

What does it do?

- All team members have a chance to give a realistic picture of what their piece of the plan requires, based on real experience
- Everyone sees why he or she is critical to the overall success of the project
- Unrealistic implementation timetables are discovered and adjusted in the planning stage
- The entire team can think creatively about how to shorten tasks that are bottlenecks
- The entire team can focus its attention and scarce resources on the truly critical tasks

How do I do it?

1. **Assemble the right team of people with firsthand knowledge of the subtasks**

2. **Brainstorm or document all the tasks needed to complete a project. Record on Post-its™**

3. Find the first task that must be done, and place the card on the extreme left of a large work surface

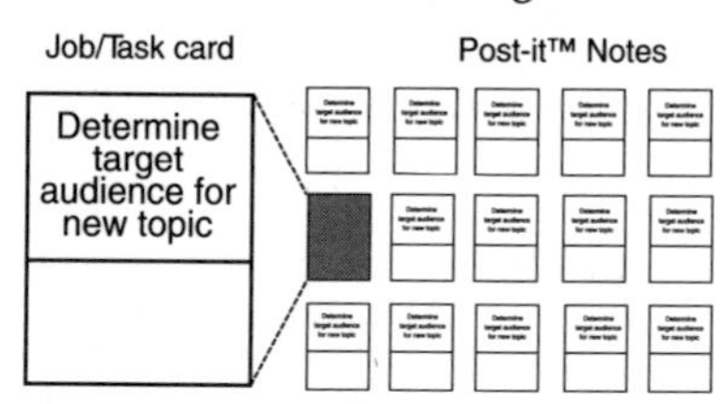

4. Ask: "Are there any tasks that can be done simultaneously with task #1?"

- If there are simultaneous tasks, place the task card above or below task #1. If not, go to the next step.

5. Ask, "What is the next task that must be done? Can others be done simultaneously?"

- Repeat this questioning process until all the recorded tasks are placed in sequence and in parallel.

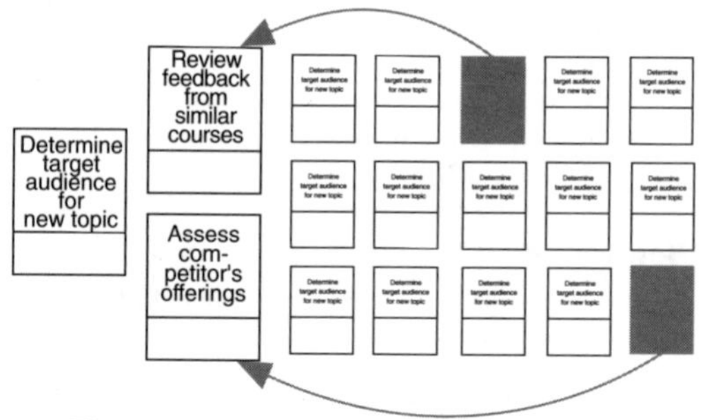

Tip At each step always ask, "Have we forgotten any other needed tasks that could be done simultaneously?"

6. **Number each task, draw the connecting arrows, and agree on a realistic time for the completion of each task**

   - Record the completion time on the bottom half of each card.

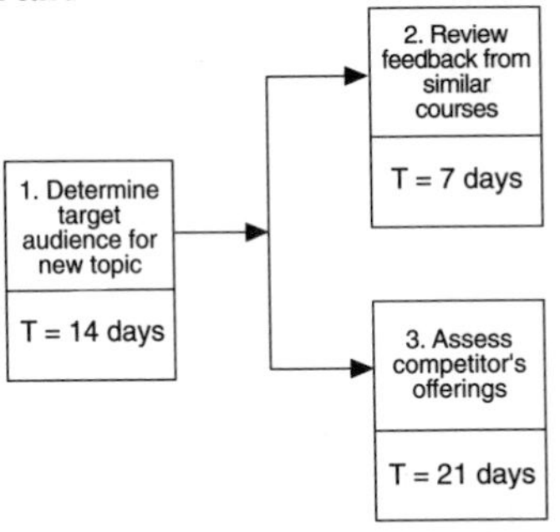

Tip Be sure to agree on the standard time unit for each task, e.g., days, weeks. Elapsed time is easier than "dedicated" time, e.g., 8 hours of dedicated time versus 8 hours over a two-week period (elapsed time).

7. **Determine the project's critical path**

   - Any delay to a task on the *critical path* will be added to the project's completion time, unless another task is accelerated or eliminated. Likewise, the project's completion time can be reduced by accelerating any task on the critical path.
   - There are two options for calculating the total critical path and the tasks included within it.

     Longest cumulative path. Identify total project completion time. Add up each path of connected activities. The longest cumulative path is the quickest possible implementation time. This is the project's *critical path.*

Calculated slack. Calculate the "slack" in the starting and completion times of each task. This identifies which tasks must be completed exactly as scheduled (on the *critical path*) and those that have some latitude.

Finding the critical path by calculating the slack

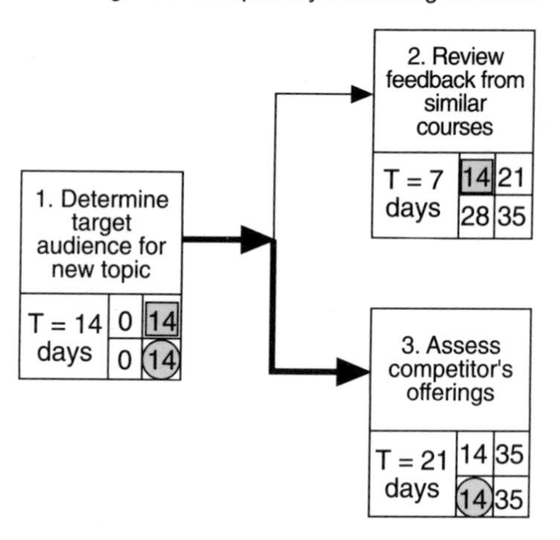

When ES = LS AND EF = LF, that task is on the critical path, and therefore there is no schedule flexibility in this task.

Tip Determining the longest cumulative path is simpler than calculating the slack, but can quickly become confusing in larger ANDs.

The calculated slack option determines the total project and slack times; and therefore the total project time and critical path are identified "automatically."

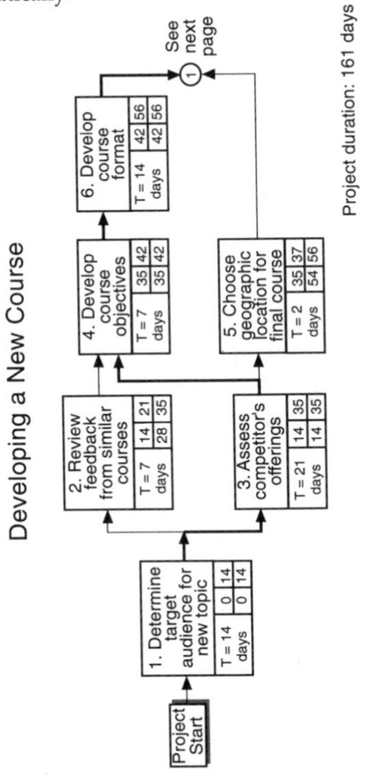

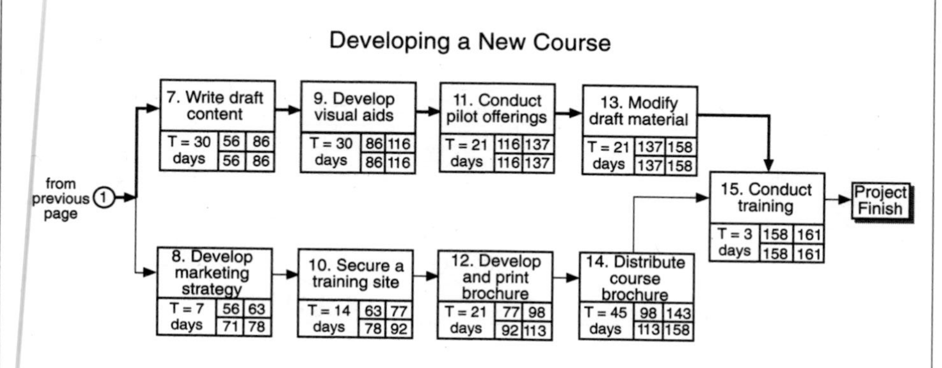
Developing a New Course
from previous page
1
7. Write draft content
T = 30 days
56 86
56 86
9. Develop visual aids
T = 30 days
86 116
86 116
11. Conduct pilot offerings
T = 21 days
116 137
116 137
13. Modify draft material
T = 21 days
137 158
137 158
15. Conduct training
T = 3 days
158 161
158 161
Project Finish
8. Develop marketing strategy
T = 7 days
56 63
71 78
10. Secure a training site
T = 14 days
63 77
78 92
12. Develop and print brochure
T = 21 days
77 98
92 113
14. Distribute course brochure
T = 45 days
98 143
113 158

Variations

The constructed example shown in this section is in the Activity on Node (AON) format. For more information on other formats such as Activity on Arrow (AOA) and Precedence Diagram (PDM), see *The Memory Jogger Plus+®*.

Another widely used, schedule-monitoring method is the Gantt chart. It is a simple tool that uses horizontal bars to show which tasks can be done simultaneously over the life of the project. Its primary disadvantage is that it cannot show which tasks are specifically dependent on each other.

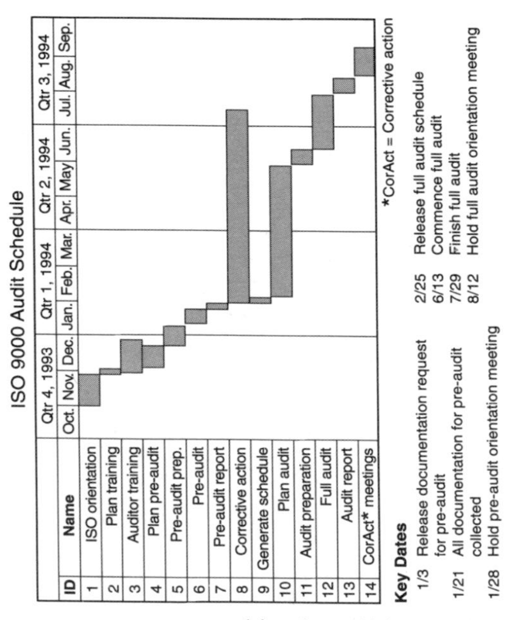

Information provided courtesy of BGP

Activity Network

Phase I
ISO 9000 Certification Audit Schedule

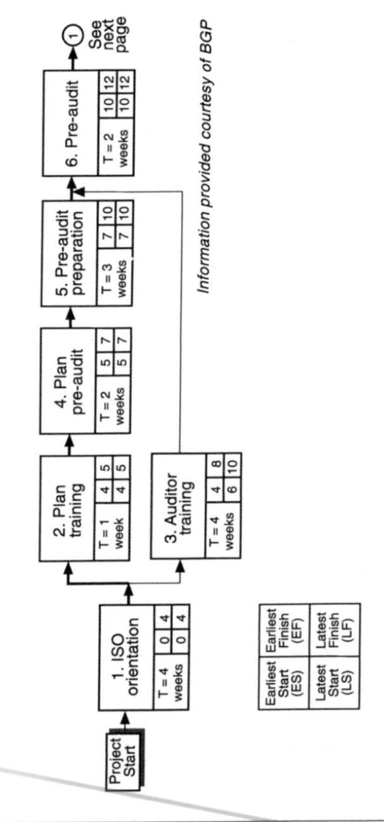

Activity Network

Phase I

ISO 9000 Certification Audit Schedule (continued)

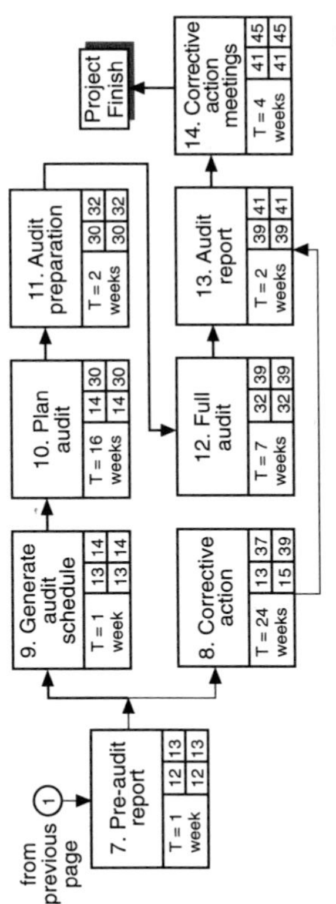

Information provided courtesy of BGP

Note: The AND shows that the certification process will take 45 weeks. The bold arrow, indicating the critical path, clearly shows those tasks that must be completed as scheduled. The tasks off the critical path will also require careful monitoring since there is only two weeks of slack time in the schedule.

Affinity Diagram

Gathering & grouping ideas

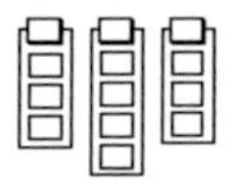

Why use it?

To allow a team to creatively generate a large number of ideas/issues and then organize and summarize natural groupings among them to understand the essence of a problem and breakthrough solutions.

What does it do?

- Encourages creativity by everyone on the team at all phases of the process
- Breaks down longstanding communication barriers
- Encourages non-traditional connections among ideas/issues
- Allows breakthroughs to emerge naturally, even on long-standing issues
- Encourages "ownership" of results that emerge because the team creates both the detailed input and general results
- Overcomes "team paralysis," which is brought on by an overwhelming array of options and lack of consensus

How do I do it?

1. **Phrase the issue under discussion in a full sentence**

> What are the issues involved in planning fun family vacations?

Tip From the start, reach consensus on the choice of words you will use. Neutral statements work well, but positive, negative, and solution-oriented questions also work.

2. **Brainstorm at least 20 ideas or issues**

   a) Follow guidelines for brainstorming.

   b) Record each idea on a Post-it™ note in bold, large print to make it visible 4–6 feet away. Use at minimum, a noun and a verb. Avoid using single words. Four to seven words work well.

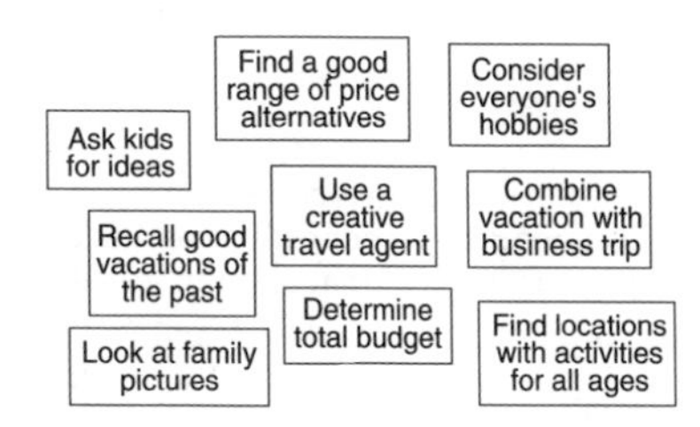

Illustration Note: There are 10 to 40 more ideas in a typical Affinity Diagram

Tip A "typical" Affinity has 40–60 items; it is not unusual to have 100–200 ideas.

3. **Without talking: sort ideas simultaneously into 5–10 related groupings**

   a) Move Post-it™ notes where they fit best for you; don't ask, simply move any notes that you think belong in another grouping.

b) Sorting will slow down or stop when each person feels sufficiently comfortable with the groupings.

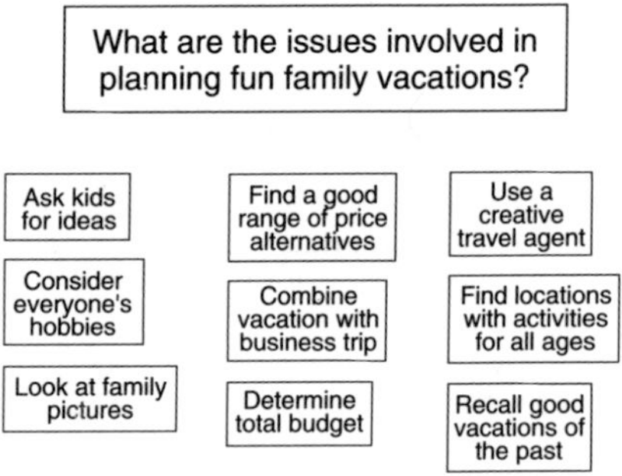

What are the issues involved in planning fun family vacations?

Ask kids for ideas	Find a good range of price alternatives	Use a creative travel agent
Consider everyone's hobbies	Combine vacation with business trip	Find locations with activities for all ages
Look at family pictures	Determine total budget	Recall good vacations of the past

Illustration Note: There are 5 to 10 more groupings of ideas in a typical Affinity Diagram

Tip Sort in silence to focus on the meaning behind and connections among all ideas, instead of emotions and "history" that often arise in discussions.

Tip As an idea is moved back and forth, try to see the logical connection that the other person is making. If this movement continues beyond a reasonable point, agree to create a duplicate Post-it™.

Tip It is okay for some notes to stand alone. These "loners" can be as important as others that fit into groupings naturally.

4. For each grouping, create summary or header cards using consensus

a) Gain a quick team consensus on a word or phrase that captures the central idea/theme of each grouping; record it on a Post-it™ note and place it at the top of each grouping. These are *draft* header cards.

b) For each grouping, agree on a concise sentence that combines the grouping's central idea and what all of the specific Post-it™ notes add to that idea; record it and replace the draft version. This is a final header card.

c) Divide large groupings into subgroups as needed and create appropriate subheaders.

d) Draw the final Affinity Diagram connecting all finalized header cards with their groupings.

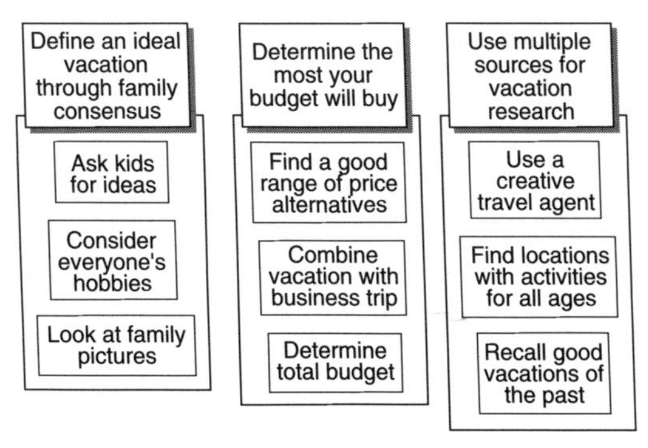

Illustration Note: There are 5 to 10 groupings of ideas in a typical Affinity. This is a partial Affinity.

Tip Spend the extra time needed to do solid header cards. Strive to capture the essence of *all* of the ideas in each grouping. ***Shortcuts here can greatly reduce the effectiveness of the final Affinity Diagram.***

It is possible that a note within a grouping could become a header card. However, don't choose the "closest one" because it's convenient. The hard work of creating new header cards often leads to breakthrough ideas.

Variations

Another popular form of this tool, called the KJ Method, was developed by the Japanese anthropologist Jiro Kawakita while he was doing fieldwork in the 1950s. The KJ Method, identified with Kawakita's initials, helped the anthropologist and his students gather and analyze data. The KJ Method differs from the Affinity Diagram described above in that the cards are fact-based and go through a highly structured refinement process before the final diagram is created.

Issues Surrounding Implementation of the Business Plan

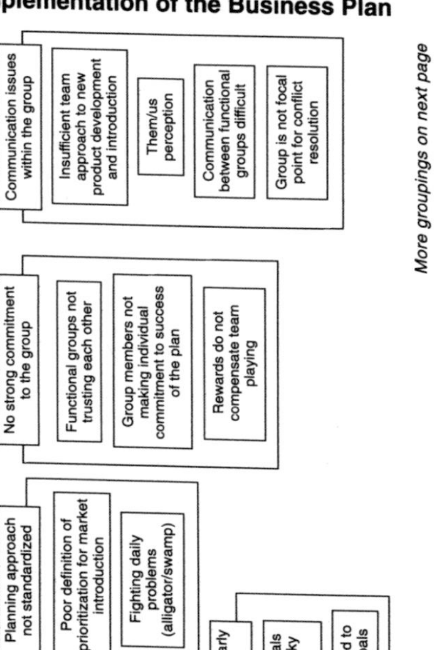

Information provided courtesy of Goodyear

More groupings on next page

Affinity

Issues Surrounding Implementation of the Business Plan (cont.)

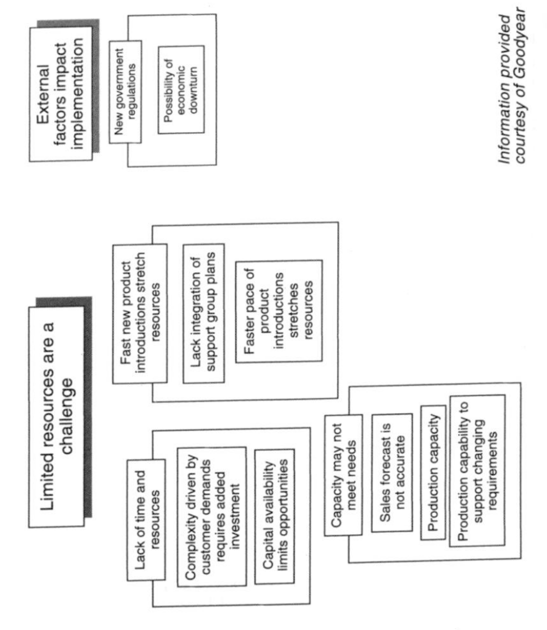

Note: The Affinity helped the team bring focus to the *many* opinions on business planning. The headers that surfaced became the key issues in the ID example (shown in the ID tool section).

Brainstorming

Creating bigger & better ideas

Why use it?

To establish a common method for a team to creatively and efficiently generate a high volume of ideas on any topic by creating a process that is free of criticism and judgment.

What does it do?

- Encourages open thinking when a team is stuck in "same old way" thinking
- Gets all team members involved and enthusiastic so that a few people don't dominate the whole group
- Allows team members to build on each other's creativity while staying focused on their joint mission

How do I do it?

There are two major methods for brainstorming.

- *Structured.* A process in which each team member gives ideas in turn.
- *Unstructured.* A process in which team members give ideas as they come to mind.

Either method can be done silently or aloud.

Structured

1. **The central brainstorming question is stated, agreed on, and written down for everyone to see**

   Be sure that everyone understands the question, issue, or problem. Check this by asking one or two members to paraphrase it before recording it on a flipchart or board.

2. **Each team member, in turn, gives an idea. No idea is criticized. Ever!**

   With each rotation around the team, any member can pass at any time. While this rotation process encourages full participation, it may also heighten anxiety for inexperienced or shy team members.

3. **As ideas are generated, write each one in large, visible letters on a flipchart or other writing surface**

   Make sure every idea is recorded with the same words of the speaker, don't interpret or abbreviate. To ensure this, the person writing should always ask the speaker if the idea has been worded accurately.

4. **Ideas are generated in turn until each person passes, indicating that the ideas (or members) are exhausted**

   Keep the process moving and relatively short—5 to 20 minutes works well, depending on how complex the topic is.

5. **Review the written list of ideas for clarity and to discard any duplicates**

   Discard only ideas that are virtually identical. It is often important to preserve subtle differences that are revealed in slightly different wordings.

Unstructured

The process is the same as in the structured method except that ideas are given by everyone at any time. There is no need to "pass" since ideas are not solicited in rotation.

Variations

There are many ways to stimulate creative team thinking. The common theme among all of them is the stimulation of creativity by taking advantage of the combined brain power of a team. Here are three examples:

- **Visual brainstorming.** Individuals (or the team) produce a picture of how they see a situation or problem.
- **Analogies/free-word association.** Unusual connections are made by comparing the problem to seemingly unrelated objects, creatures, or words. For example: "If the problem was an animal, what kind would it be?"
- **6-3-5 method.** This powerful, silent method is proposed by Helmut Schlicksupp in his book *Creativity Workshop*. It is done as follows:

  a) Based on a single brainstorming issue, each person on the team (usually 6 people) has 5 minutes to write down 3 ideas on a sheet of paper.

  b) Each person then passes his or her sheet of paper to the next person, who has 5 more minutes to add 3 more ideas that build on the first 3 ideas.

c) This rotation is repeated as many times as there are team members, e.g., 6 team members = 6 rotations, 6 sheets of paper, 18 ideas per sheet.

This interesting process forces team members to consciously build on each other's perspectives and input.

Cause & Effect/ Fishbone Diagram

Find & cure causes, NOT symptoms

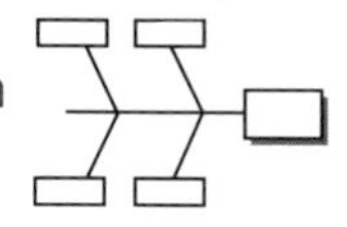

Why use it?

To allow a team to identify, explore, and graphically display, in increasing detail, all of the possible causes related to a problem or condition to discover its root cause(s).

What does it do?

- Enables a team to focus on the content of the problem, not on the history of the problem or differing personal interests of team members
- Creates a snapshot of the collective knowledge and consensus of a team around a problem. This builds support for the resulting solutions
- Focuses the team on causes, not symptoms

How do I do it?

1. **Select the most appropriate cause & effect format. There are two major formats:**
   - **Dispersion Analysis Type** is constructed by placing individual causes within each "major" cause category and then asking of each individual cause "Why does this cause (dispersion) happen?" This question is repeated for the next level of detail until the team runs out of causes. The graphic examples shown in Step 3 of this tool section are based on this format.

• **Process Classification Type** uses the major steps of the process in place of the major cause categories. The root cause questioning process is the same as the Dispersion Analysis Type.

2. **Generate the causes needed to build a Cause & Effect Diagram. Choose one method:**
   - **Brainstorming** without previous preparation
   - **Check Sheets** based on data collected by team members before the meeting

3. **Construct the Cause & Effect/Fishbone Diagram**

   a) Place the problem statement in a box on the righthand side of the writing surface.

   - Allow plenty of space. Use a flipchart sheet, butcher paper, or a large white board. A paper surface is preferred since the final Cause & Effect Diagram can be moved.

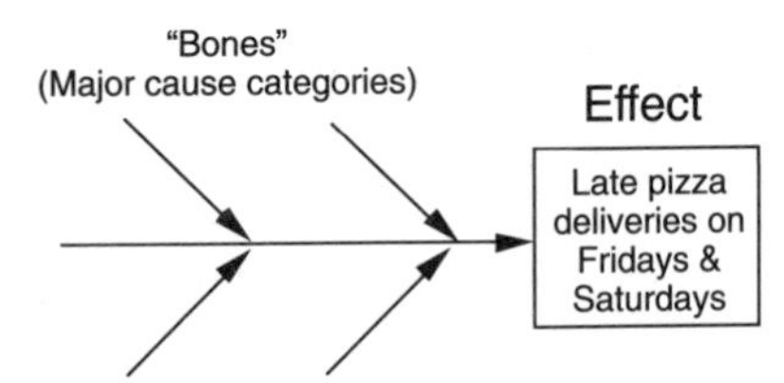

Tip Make sure everyone agrees on the problem statement. Include as much information as possible on the "what," "where," "when," and "how much" of the problem. Use data to specify the problem.

b) Draw major cause categories or steps in the production or service process. Connect them to the "backbone" of the fishbone chart.

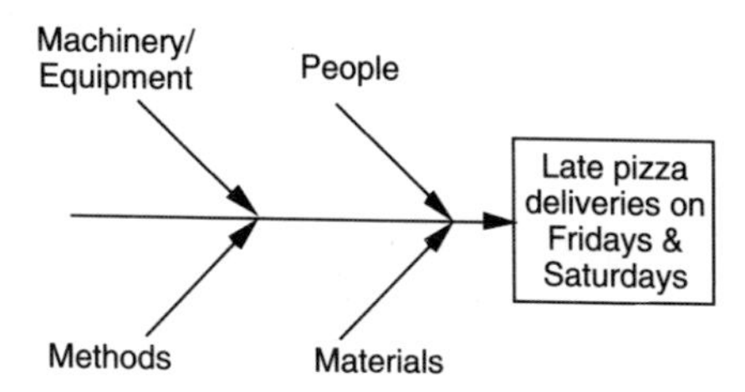

Illustration Note: In a Process Classification Type format, replace the major "bone" categories with: "Order Taking," "Preparation," "Cooking," and "Delivery."

- Be flexible in the major cause "bones" that are used. In a **Production Process** the traditional categories are: **Machines** (equipment), **Methods** (how work is done), **Materials** (components or raw materials), and **People** (the human element). In a **Service Process** the traditional methods are: **Policies** (higher-level decision rules), **Procedures** (steps in a task), **Plant** (equipment and space), and **People**. In both types of processes, **Environment** (buildings, logistics, and space), and **Measurement** (calibration and data collection) are also frequently used. *There is no perfect set or number of categories. Make them fit the problem.*

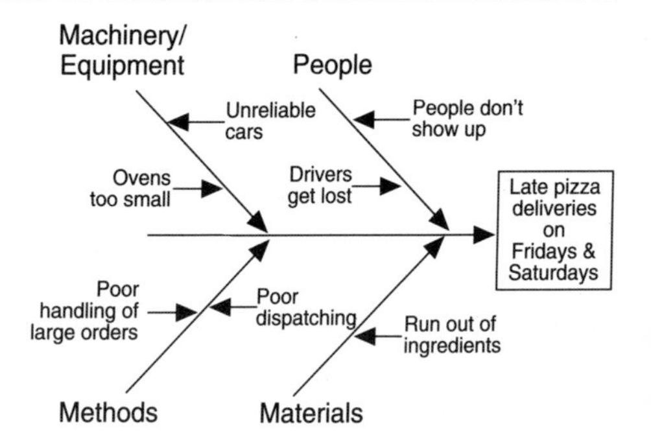

c) Place the brainstormed or data-based causes in the appropriate category.

- In brainstorming, possible causes can be placed in a major cause category as each is generated, or only after the entire list has been created. Either works well but brainstorming the whole list first maintains the creative flow of ideas without being constrained by the major cause categories or where the ideas fit in each "bone."
- Some causes seem to fit in more than one category. Ideally each cause should be in only one category, but some of the "people" causes may legitimately belong in two places. Place them in both categories and see how they work out in the end.

Tip If ideas are slow in coming, use the major cause categories as catalysts, e.g., "What in 'materials' is causing . . . ?"

d) Ask repeatedly of each cause listed on the "bones," either:

- "Why does it happen?" For example, under "Run out of ingredients" this question would lead to more basic causes such as "Inaccurate ordering," "Poor use of space," and so on.

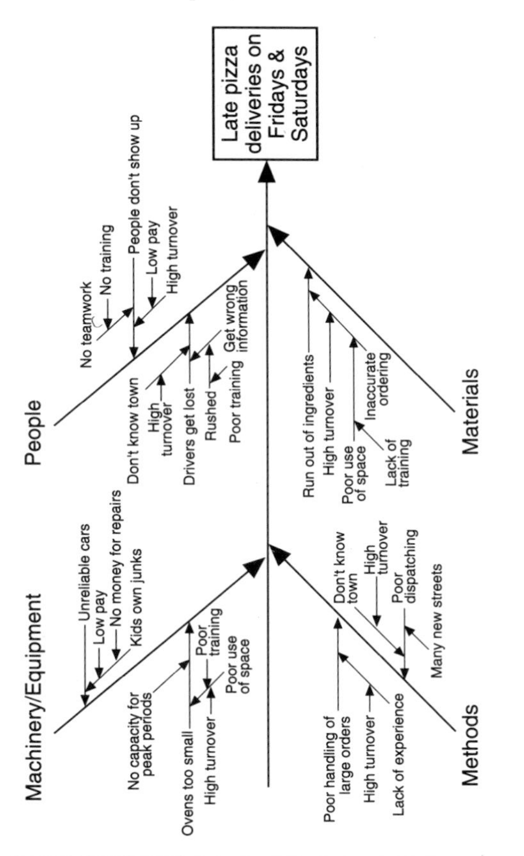

- "What could happen?" For example, under "Run out of ingredients" this question would lead to a deeper understanding of the problem such as "Boxes," "Prepared dough," "Toppings," and so on.

Tip For each deeper cause, continue to push for deeper understanding, but know when to stop. A rule of thumb is to stop questioning when a cause is controlled by more than one level of management removed from the group. Otherwise, the process could become an exercise in frustration. Use common sense.

e) Interpret or test for root cause(s) by one or more of the following:

- Look for causes that appear repeatedly within or across major cause categories.
- Select through either an unstructured consensus process or one that is structured, such as Nominal Group Technique or Multivoting.
- Gather data through Check Sheets or other formats to determine the relative frequencies of the different causes.

Variations

Traditionally, Cause & Effect Diagrams have been created in a meeting setting. The completed "fishbone" is often reviewed by others and/or confirmed with data collection. A very effective alternative is CEDAC®, in which a large, highly visible, blank fishbone chart is displayed prominently in a work area. Everyone posts both potential causes and solutions on Post-it™ notes in each of the categories. Causes and solutions are reviewed, tested, and posted. This technique opens up the process to the knowledge and creativity of every person in the operation.

Cause & Effect/Fishbone

Bed Assignment Delay

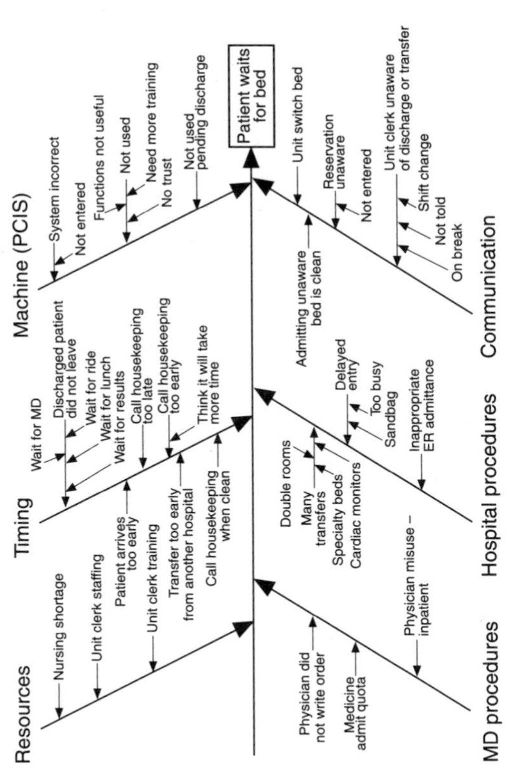

Information provided courtesy of Rush-Presbyterian-St. Luke's Medical Center

Cause & Effect/Fishbone

Causes for Bent Pins (Plug-In Side)

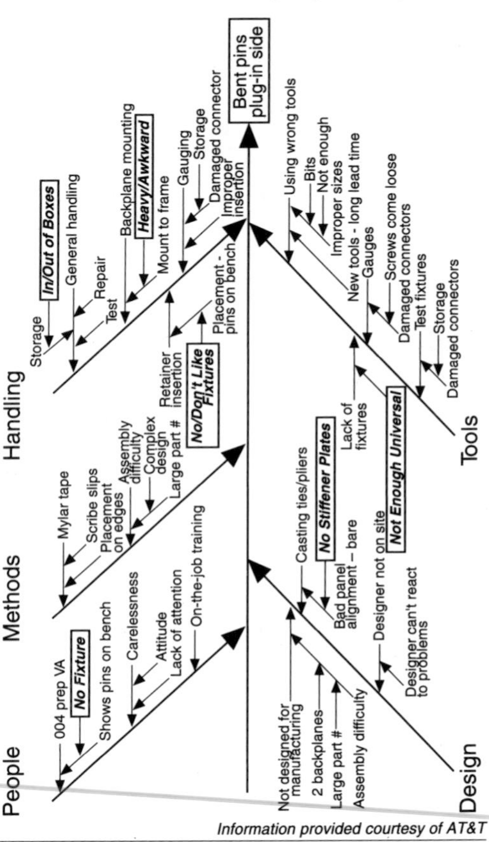

Information provided courtesy of AT&T

Check Sheet

Counting & accumulating data

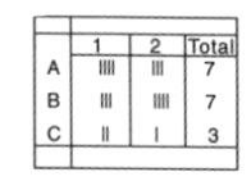

	1	2	Total
A	IIII	III	7
B	III	IIII	7
C	II	I	3

Why use it?

To allow a team to systematically record and compile data from historical sources, or observations as they happen, so that patterns and trends can be clearly detected and shown.

What does it do?

- Creates easy-to-understand data that come from a simple, efficient process that can be applied to any key performance areas
- Builds, with each observation, a clearer picture of "the facts" as opposed to the opinions of each team member
- Forces agreement on the definition of each condition or event (every person has to be looking for and recording the same thing)
- Makes patterns in the data become obvious quickly

How do I do it?

1. **Agree on the definition of the events or conditions being observed**
   - If you are building a list of events or conditions as the observations are made, agree on the overall definition of the project.

     Example: If you are looking for reasons for late payments, agree on the definition of "late."

- If you are working from a standard list of events or conditions, make sure that there is agreement on the meaning and application of each one.

  Example: If you are tracking sales calls from various regions, make sure everyone knows which states are in each region.

2. **Decide who will collect the data; over what period; and from what sources**

   - Who collects the data obviously depends on the project and resources. The data collection period can range from hours to months. The data can come from either a sample or an entire population.
   - Make sure the data collector(s) have both the time and knowledge they need to collect accurate information.
   - Collect the data over a sufficient period to be sure the data represents "typical" results during a "typical" cycle for your business.
   - Sometimes there may be important differences within a population that should be reflected by sampling each different subgroup individually. This is called stratification.

     Example: Collect complaint data from business travelers separately from other types of travelers. Collect scrap data from each machine separately.

   Tip It must be safe to record and report "bad news," otherwise the data will be filtered.

3. **Design a Check Sheet form that is clear, complete, and easy to use**
   - A complete Check Sheet, illustrated below, includes the following:

     Source Information

     a Name of project
     b Location of data collection
     c Name of person recording data, if it applies
     d Date(s)
     e Other important identifiers

     Content Information

     f Column with defect/event name
     g Columns with collection days/dates
     h Totals for each column
     i Totals for each row
     j Grand total for both the columns and rows

a Project: Admission Delays			c Name: (if applicable)				e Shift: All	
b Location: Emergency Room			d Dates: 3/10 to 3/16					
f Reason:	g Date							i Total
	3/10	3/11	3/12	3/13	3/14	3/15	3/16	
Lab delays	9	4	6	6	3	12	12	52
No beds available	2	7	2	4	5	8	3	31
Incomplete patient information	7	3	1	2	2	4	5	24
h Total	33	28	36	30	25	47	38	j 237

4. **Collect the data consistently and accurately**
   - Make sure all entries are written clearly.

   Tip Managers and/or team members can do their part to help the data collector(s) do their job well by simply showing an interest in the project. Ask the collector(s) how the project is working out. Show your support—tell the data collector(s) you think it is important to collect the information. ***Above all—act on the data as quickly as possible!***

Variations

Defect Location

Shows the concentration of defects by marking a drawing of the product each time a defect is found.

Project: Monitor casing damage
Location: Assembly
Name: (if applicable)
Date: 6/11
Shift: First

Front

Back

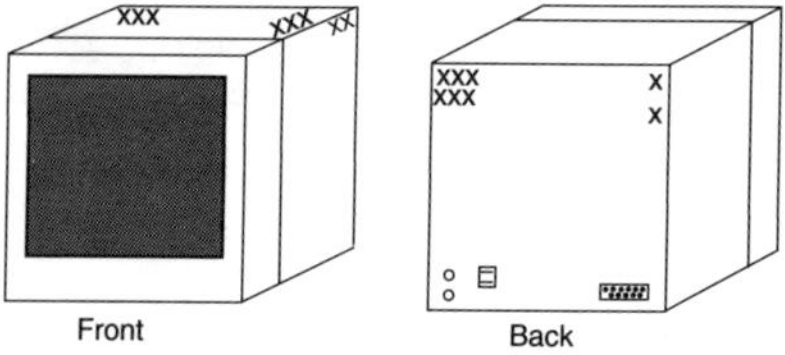

Generic computer monitor

Task Checklist

Tasks in producing a product or delivering a service are checked off as they are done. In complex processes this is a form of "mistake-proofing."

Check Sheet

Keyboard Errors in Class Assignment

Mistakes	March			
	1	2	3	Total
Centering	II	III	III	8
Spelling	卌 II	卌 卌 I	卌	23
Punctuation	卌 卌 卌	卌 卌	卌 卌 卌	40
Missed paragraph	II	I	I	4
Wrong numbers	III	IIII	III	10
Wrong page numbers	I	I	II	4
Tables	IIII	卌	IIII	13
Total	34	35	33	102

Information provided courtesy of Millcreek Township School District, Millcreek Township, Pennsylvania

Control Charts

Recognizing sources of variation

Why use it?

To monitor, control, and improve process performance over time by studying variation and its source.

What does it do?

- Focuses attention on detecting and monitoring process variation over time
- Distinguishes special from common causes of variation, as a guide to local or management action
- Serves as a tool for ongoing control of a process
- Helps improve a process to perform consistently and predictably for higher quality, lower cost, and higher effective capacity
- Provides a common language for discussing process performance

How do I do it?

There are many types of Control Charts. The Control Chart(s) that your team decides to use will be determined by the type of data you have. Use the Tree Diagram on the next page to determine which Control Chart(s) will best fit your situation. Other types of Control Charts, which are beyond the scope of this book, include the Pre-Control Chart, the Moving Average & Range Chart, the Cumulative Sum Chart, and Box Plots.

Based on the type of data and sample size you have, choose the appropriate Control Chart.

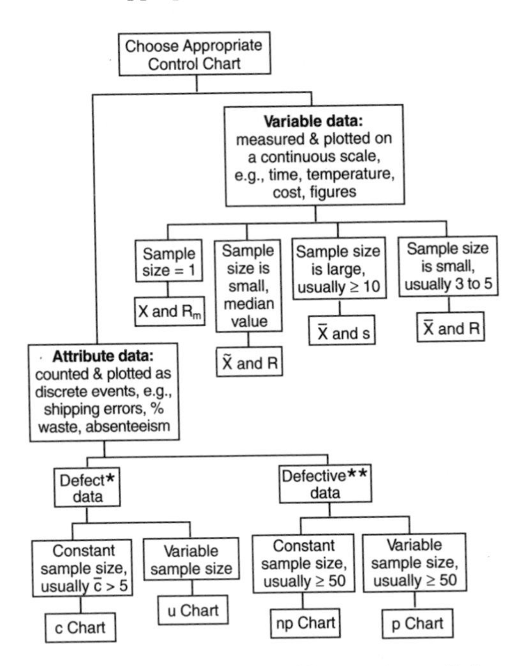

* Defect = Failure to meet one of the acceptance criteria. A defective unit might have multiple defects.

** Defective = An entire unit fails to meet acceptance criteria, regardless of the number of defects on the unit.

Constructing Control Charts

1. **Select the process to be charted**

2. **Determine sampling method and plan**
   - How large a sample can be drawn? Balance the time and cost to collect a sample with the amount of information you will gather. *See the Tree Diagram on the previous page for suggested sample sizes.*
   - As much as possible, obtain the samples under the same technical conditions: the same machine, operator, lot, and so on.
   - Frequency of sampling will depend on whether you are able to discern patterns in the data. Consider hourly, daily, shifts, monthly, annually, lots, and so on. Once the process is "in control," you might consider reducing the frequency with which you sample.
   - Generally, collect 20–25 groups of samples before calculating the statistics and control limits.
   - Consider using historical data to set a baseline.

   Tip Make sure samples are random. To establish the inherent variation of a process, allow the process to run untouched, i.e., according to standard procedures.

3. **Initiate data collection**
   - Run the process untouched, and gather sampled data.
   - Record data on an appropriate Control Chart sheet or other graph paper. Include any unusual events that occur.

4. Calculate the appropriate statistics

a) If you have attribute data, use the Attribute Data Table, Central Line column.

Attribute Data Table

Type Control Chart	Sample size	Central Line	Control Limits
Fraction defective p Chart	Variable, usually ≥50	For each subgroup: $p = np/n$ For all subgroups: $\bar{p} = \Sigma np/\Sigma n$	$^*UCL_p = \bar{p} + 3\sqrt{\frac{\bar{p}(1-\bar{p})}{n}}$ $^*LCL_p = \bar{p} - 3\sqrt{\frac{\bar{p}(1-\bar{p})}{n}}$
Number defective np Chart	Constant, usually ≥50	For each subgroup: np = # defective units For all subgroups: $n\bar{p} = \Sigma np/k$	$UCL_{np} = n\bar{p} + 3\sqrt{n\bar{p}(1-\bar{p})}$ $LCL_{np} = n\bar{p} - 3\sqrt{n\bar{p}(1-\bar{p})}$
Number of defects c Chart	Constant	For each subgroup: c = # defects For all subgroups: $\bar{c} = \Sigma c/k$	$UCL_c = \bar{c} + 3\sqrt{\bar{c}}$ $LCL_c = \bar{c} - 3\sqrt{\bar{c}}$
Number of defects per unit u Chart	Variable	For each subgroup: $u = c/n$ For all subgroups: $\bar{u} = \Sigma c/\Sigma n$	$^*UCL_u = \bar{u} + 3\sqrt{\frac{\bar{u}}{n}}$ $^*LCL_u = \bar{u} - 3\sqrt{\frac{\bar{u}}{n}}$

np = # defective units
c = # of defects
n = sample size within each subgroup
k = # of subgroups

* This formula creates changing control limits. To avoid this, use average sample sizes $\bar{n}$ for those samples that are within ±20% of the average sample size. Calculate individual limits for the samples exceeding ±20%.

b) If you have variable data, use the Variable Data Table, Central Line column.

Variable Data Table

Type Control Chart	Sample size n	Central Line*	Control Limits
Average & Range	<10, but usually 3 to 5	$\bar{\bar{X}} = \frac{(\bar{X}_1+\bar{X}_2+\ldots\bar{X}_k)}{k}$	$UCL_{\bar{X}} = \bar{\bar{X}} + A_2\bar{R}$ $LCL_{\bar{X}} = \bar{\bar{X}} - A_2\bar{R}$
$\bar{X}$ and R		$\bar{R} = \frac{(R_1+R_2+\ldots R_k)}{k}$	$UCL_R = D_4\bar{R}$ $LCL_R = D_3\bar{R}$
Average & Standard Deviation	Usually ≥10	$\bar{\bar{X}} = \frac{(\bar{X}_1+\bar{X}_2+\ldots\bar{X}_k)}{k}$	$UCL_{\bar{X}} = \bar{\bar{X}} + A_3\bar{s}$ $LCL_{\bar{X}} = \bar{\bar{X}} - A_3\bar{s}$
$\bar{X}$ and s		$\bar{s} = \frac{(s_1+s_2+\ldots s_k)}{k}$	$UCL_s = B_4\bar{s}$ $LCL_s = B_3\bar{s}$
Median & Range	<10, but usually 3 or 5	$\bar{\tilde{X}} = \frac{(\tilde{X}_1+\tilde{X}_2+\ldots\tilde{X}_k)}{k}$	$UCL_{\tilde{X}} = \bar{\tilde{X}} + \tilde{A}_2\bar{R}$ $LCL_{\tilde{X}} = \bar{\tilde{X}} - \tilde{A}_2\bar{R}$
$\tilde{X}$ and R		$\bar{R} = \frac{(R_1+R_2+\ldots R_k)}{k}$	$UCL_R = D_4\bar{R}$ $LCL_R = D_3\bar{R}$
Individuals & Moving Range	1	$\bar{X} = \frac{(X_1+X_2+\ldots X_k)}{k}$	$UCL_X = \bar{X} + E_2\bar{R}_m$ $LCL_X = \bar{X} - E_2\bar{R}_m$
X and R_m		$R_m = \lvert(X_{i+1} - X_i)\rvert$ $\bar{R}_m = \frac{(R_1 + R_2 + \ldots R_{k-1})}{k-1}$	$UCL_{Rm} = D_4\bar{R}_m$ $LCL_{Rm} = D_3\bar{R}_m$

k = # of subgroups, $\tilde{X}$ = median value within each subgroup

$^*\bar{X} = \frac{\Sigma X_i}{n}$

5. **Calculate the control limits**

   a) If you have attribute data, use the Attribute Data Table, Control Limits column.

   b) If you have variable data, use the Variable Data Table, Control Limits column for the correct formula to use.

   - Use the Table of Constants to match the numeric values to the constants in the formulas shown in the Control Limits column of the Variable Data Table. The values you will need to look up will depend on the type of Variable Control Chart you choose and on the size of the sample you have drawn.

   Tip If the Lower Control Limit (LCL) of an Attribute Data Control Chart is a negative number, set the LCL to zero.

   Tip The p and u formulas create changing control limits if the sample sizes vary subgroup to subgroup. To avoid this, use the average sample size, $\bar{n}$, for those samples that are within ±20% of the average sample size. Calculate individual limits for the samples exceeding ±20%.

6. **Construct the Control Chart(s)**

   - For Attribute Data Control Charts, construct one chart, plotting each subgroup's proportion or number defective, number of defects, or defects per unit.
   - For Variable Data Control Charts, construct two charts: on the top chart plot each subgroup's mean, median, or individuals, and on the bottom chart plot each subgroup's range or standard deviation.

Table of Constants

Sample size n	$\bar{X}$ and R Chart			$\bar{X}$ and s Chart			
	A_2	D_3	D_4	A_3	B_3	B_4	c_4*
2	1.880	0	3.267	2.659	0	3.267	.7979
3	1.023	0	2.574	1.954	0	2.568	.8862
4	0.729	0	2.282	1.628	0	2.266	.9213
5	0.577	0	2.114	1.427	0	2.089	.9400
6	0.483	0	2.004	1.287	0.030	1.970	.9515
7	0.419	0.076	1.924	1.182	0.118	1.882	.9594
8	0.373	0.136	1.864	1.099	0.185	1.815	.9650
9	0.337	0.184	1.816	1.032	0.239	1.761	.9693
10	0.308	0.223	1.777	0.975	0.284	1.716	.9727

Sample Size n	$\tilde{X}$ and R Chart			X and R_m Chart			
	$\tilde{A}_2$	D_3	D_4	E_2	D_3	D_4	d_2*
2	----	0	3.267	2.659	0	3.267	1.128
3	1.187	0	2.574	1.772	0	2.574	1.693
4	----	0	2.282	1.457	0	2.282	2.059
5	0.691	0	2.114	1.290	0	2.114	2.326
6	----	0	2.004	1.184	0	2.004	2.534
7	0.509	0.076	1.924	1.109	0.076	1.924	2.704
8	----	0.136	1.864	1.054	0.136	1.864	2.847
9	0.412	0.184	1.816	1.010	0.184	1.816	2.970
10	----	0.223	1.777	0.975	0.223	1.777	3.078

* Useful in estimating the process standard deviation $\hat{\sigma}$.

Note: The minimum sample size shown in this chart is 2 because variation in the form of a range can only be calculated in samples greater than 1. The X and R_m Chart creates these minimum samples by combining and then calculating the difference between sequential, individual measurements.

- Draw a solid horizontal line on each chart. This line corresponds to the process average.
- Draw dashed lines for the upper and lower control limits.

Interpreting Control Charts

- **Attribute Data Control Charts** are based on one chart. The charts for fraction or number defective, number of defects, or number of defects per unit, measure variation *between samples*. **Variable Data Control Charts** are based on two charts: the one on top, for averages, medians, and individuals, measures variation *between subgroups* over time; the chart below, for ranges and standard deviations, measures variation *within subgroups* over time.
- Determine if the process mean (center line) is where it should be relative to your customer specifications or your internal business needs or objectives. If not, then it is an indication that something has changed in the process, or the customer requirements or objectives have changed.
- Analyze the data relative to the control limits; distinguishing between *common* causes and *special* causes. The fluctuation of the points within the limits results from variation inherent in the process. This variation results from common causes within the system, e.g., design, choice of machine, preventive maintenance, and can only be affected by changing that system. However, points outside of the limits or patterns within the limits, come from a special cause, e.g., human errors, unplanned events, freak occurrences, that is not part of the way the process normally operates, or is present because of an unlikely combination of process steps. Special causes must

be eliminated before the Control Chart can be used as a monitoring tool. Once this is done, the process will be "in control" and samples can be taken at regular intervals to make sure that the process doesn't fundamentally change. See "Determining if Your Process is Out of Control."

- Your process is in "statistical control" if the process is not being affected by special causes, the influence of an individual or machine. All the points must fall within the control limits and they must be randomly dispersed about the average line for an in-control system.

Tip "Control" doesn't necessarily mean that the product or service will meet your needs. It only means that the process is *consistent.* Don't confuse control limits with specification limits—specification limits are related to customer requirements, not process variation.

Tip Any points outside the control limits, once identified with a cause (or causes), should be removed and the calculations and charts redone. Points within the control limits, but showing indications of trends, shifts, or instability, are also special causes.

Tip When a Control Chart has been initiated and all special causes removed, continue to plot new data on a new chart, but DO NOT recalculate the control limits. As long as the process does not change, the limits should not be changed. Control limits should be recalculated only when a permanent, desired change has occurred in the process, and only using data *after* the change occurred.

Tip Nothing will change just because you charted it! You need to do something. Form a team to investigate. See "Common Questions for Investigating an Out-of-Control Process."

Determining if Your Process is "Out of Control"

A process is said to be "out of control" if either one of these is true:

1. **One or more points fall outside of the control limits**

2. **When the Control Chart is divided into zones, as shown below, any of the following points are true:**

Zone	
- - - - - - - - - - -	Upper Control Limit (UCL)
Zone A	
Zone B	
Zone C	Average
Zone C	
Zone B	
Zone A	
- - - - - - - - - - -	Lower Control Limit (LCL)

a) Two points, out of three consecutive points, are on the same side of the average in Zone A or beyond.

b) Four points, out of five consecutive points, are on the same side of the average in Zone B or beyond.

c) Nine consecutive points are on one side of the average.

d) There are six consecutive points, increasing or decreasing.

e) There are fourteen consecutive points that alternate up and down.

f) There are fifteen consecutive points within Zone C (above and below the average).

Tests for Control

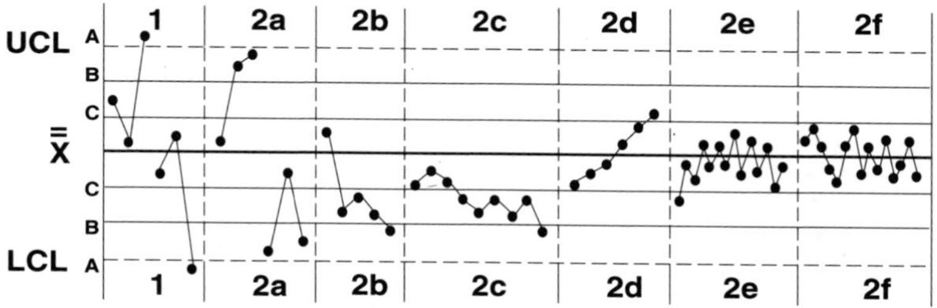

Source: Lloyd S. Nelson, Director of Statistical Methods, Nashua Corporation, New Hampshire

Common Questions for Investigating an Out-of-Control Process

❐ Yes ❐ No Are there differences in the measurement accuracy of instruments/methods used?

❐ Yes ❐ No Are there differences in the methods used by different personnel?

❐ Yes ❐ No Is the process affected by the environment, e.g., temperature, humidity?

❐ Yes ❐ No Has there been a significant change in the environment?

❐ Yes ❐ No Is the process affected by predictable conditions? Example: tool wear.

❐ Yes ❐ No Were any untrained personnel involved in the process at the time?

❐ Yes ❐ No Has there been a change in the source for input to the process? Example: raw materials, information.

❐ Yes ❐ No Is the process affected by employee fatigue?

❐ Yes ❐ No Has there been a change in policies or procedures? Example: maintenance procedures.

❐ Yes ❐ No Is the process adjusted frequently?

❐ Yes ❐ No Did the samples come from different parts of the process? Shifts? Individuals?

❐ Yes ❐ No Are employees afraid to report "bad news"?

A team should address each "Yes" answer as a potential source of a special cause.

Individuals & Moving Range Chart

IV Lines Connection Time

Process/Operation: IV Lines Connection Open Heart Admissions			**Department:** Intensive Care
Characteristic: Time in seconds	**Sample Size:** One	**Sample Frequency:** Each patient	**By:** EW **Date:** 6/10
Individuals: $k = 26$ $\sum X = 8470$	$\bar{\bar{X}} = 325.77$	**UCL** = 645	**LCL** = 7
Ranges: $n = 2$ $\sum R = 2990$	$\bar{R} = 119.6$	**UCL** = 392	**LCL** = 0

Seconds

720
630
540
450
360
270
180
90
0

UCL
$\bar{X}$
LCL

X	600	480	540	240	420	450	480	690	240	360	450	300	480	120	240	210	210	180	240	300	300	130	120	300	180	210
R_m	—	120	60	300	180	30	30	210	450	120	90	150	180	360	120	30	0	30	60	60	0	120	60	80	120	30
Who?	EW	EW	EW	EW	EW	EW	EW	EW	MA	EW	EW	EW	EW	EW	EW	EW	MA	EW	EW	EW	EW	EW	EW	EW	EW	EW
When?	4/9	11	12	13	25	30	5/2	3	4	5	9	13	14	14	16	17	19	20	22	23	23	24	30	30	6/6	6

420
360
300
240
180
120
60
0

UCL
$\bar{R}_m$
LCL

Information provided courtesy of Parkview Episcopal Medical Center

Note: Something in the process changed, and now it takes less time to make IV connections for patients being admitted for open heart surgery.

p Chart

General Dentistry: Percent of Patients Who Failed to Keep Appointments

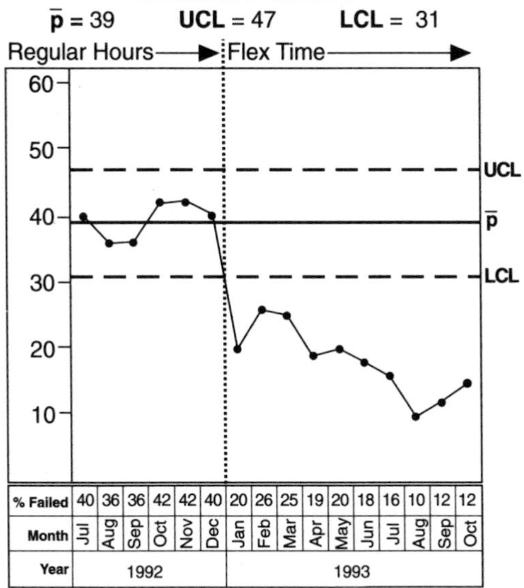

Information provided courtesy of U.S. Navy, Naval Dental Center, San Diego

Note: Providing flex time for patients resulted in fewer appointments missed.

u Chart

Shop Process Check
Solder Defects

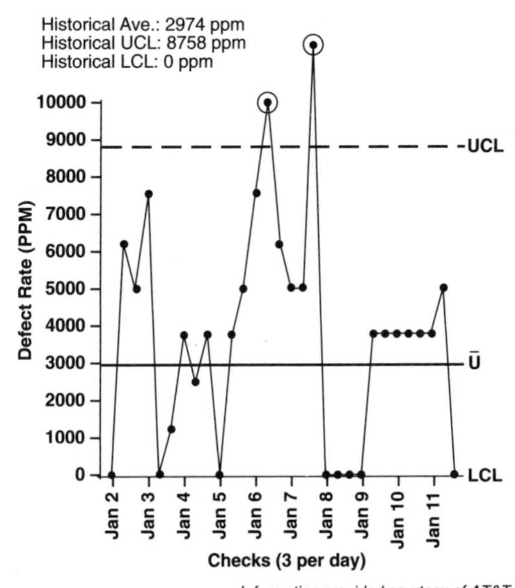

Information provided courtesy of AT&T

$\overline{X}$ & R Chart

Overall Course Evaluations

n = 10 evaluations randomly sampled each week

1-Not at all 2-Not very 3-Moderately 4-Very 5-Extremely

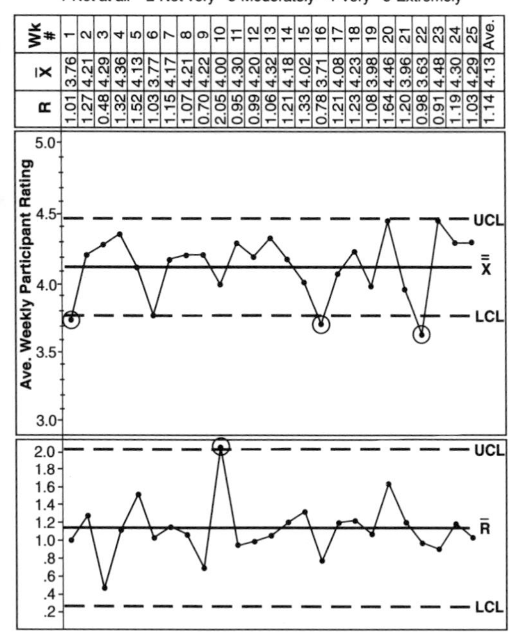

Wk #	1	2	3	4	5	6	7	8	9	10	11	12	13	14	15	16	17	18	19	20	21	22	23	24	25	Ave.
$\overline{X}$	3.76	4.21	4.29	4.36	4.13	3.77	4.17	4.21	4.22	4.00	4.30	4.20	4.32	4.18	4.02	3.71	4.08	4.23	3.98	4.46	3.96	3.63	4.48	4.30	4.29	4.13
R	1.01	1.27	0.48	1.32	1.52	1.03	1.15	1.07	0.70	2.05	0.95	0.99	1.06	1.21	1.33	0.78	1.21	1.23	1.08	1.64	1.20	0.98	0.91	1.19	1.03	1.14

Information provided courtesy of Hamilton Standard

Note: Weeks 1, 10 (from bottom chart), 16, and 22 should be reviewed to understand why the ratings are outside the control limits.

Data Points

Turning data into information

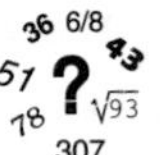

What Type of Data Do You Have?

- Words?
- Numbers?
  - *Attribute data?* Attribute data can be counted and plotted as discrete events. It includes the count of the numbers or percentages of good or bad, right or wrong, pass or fail, yes or no.

    Example: Number of correct answers on a test, number of mistakes per typed page, percent defective product per shift.
  - *Variable data?* Variable data can be measured and plotted on a continuous scale.

    Example: Length, time, volume, weight.

Do You Need to Collect Data?

- If you need to know the performance of an entire population, the more economical and less time consuming method is to draw a sample from a population. With a sample, you can make inferences about, or predict, the performance of a population. Basic sampling methods:
  - *Random.* Each and every observation or data measure has an equally likely chance of being selected. Use a random number table or random number generator to select the samples.
  - *Sequential.* Every *nth* sample is selected.
  - *Stratified.* A sample is taken from stratified data groups.

Can You Categorize Your Data Into Subgroups?

- When you stratify data, you break it down into meaningful subcategories or classifications, and from this point you can focus your problem solving.

  Example: Data often comes from many sources but is treated as if coming from one. Data on minor injuries for a plant may be recorded as a single figure, but that number is actually the sum total of injuries by 1) type (cuts, burns, scrapes), 2) location (eyes, hands, feet), and 3) department (maintenance, shipping, production). Below is an example of how data has been stratified by plant department.

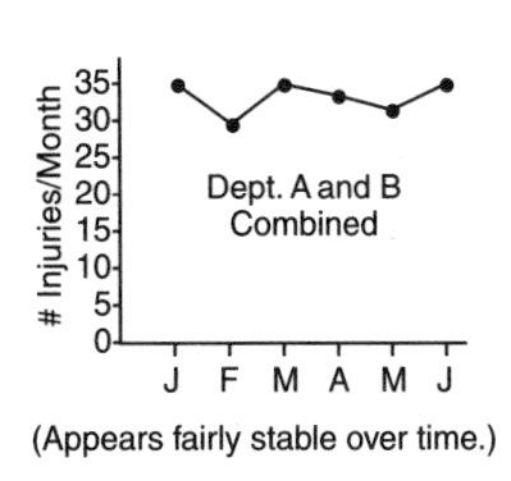

(Appears fairly stable over time.)

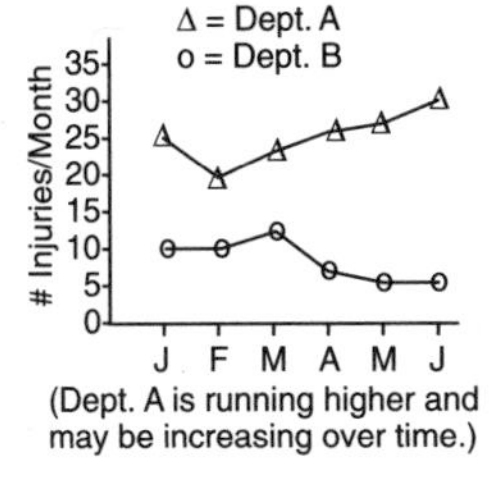

(Dept. A is running higher and may be increasing over time.)

What Patterns are Important in Your Data?

Predictable patterns or distributions can be described with statistics.

- **Measures of location**
  - *Mean* (or average). Represented by $\bar{X}$ (or X-bar), the mean is the sum of the values of the sample ($X_1, X_2, X_3 \ldots X_n$) divided by the total number (n) of sampled data.

    Example: For the sample: (3, 5, 4, 7, 5)

$$\bar{X} = \frac{(3 + 5 + 4 + 7 + 5)}{5} = 4.8$$

- *Median.* When sampled data are rank ordered, lowest to highest, the median is the middle number.

  Example: For the sample: (3, 5, 4, 7, 5) Median of (3, 4, 5, 5, 7) = 5

  When there are an even number of values, the median is the average of the middle two values.

  Example: For the sample: (2, 5, 7, 4, 5, 3) Median of (2, 3, 4, 5, 5, 7) = 4.5

- *Mode.* The most frequently occurring value(s) in a sample.

  Example: For the sample: (3, 5, 4, 7, 5) Mode = 5

- **Measures of variation**
  - *Range.* Represented by R, the range is the difference between the highest data value (X_{max}) and the lowest data value (X_{min}).

    Example: For the sample: (3, 5, 4, 7, 5) R = 7 – 3 = 4

  - *Standard Deviation.* Represented by s, the standard deviation of a sample measures the variation of the data around the mean. The less variation there is of the data values about the mean, $\bar{X}$, the closer s will be to zero (0).

Example: For the sample: (3, 5, 4, 7, 5) $\bar{X} = 4.8$

$$s = \sqrt{\frac{[(3-4.8)^2 + (5-4.8)^2 + (4-4.8)^2 + (7-4.8)^2 + (5-4.8)^2]}{5-1}}$$

$$= \sqrt{\frac{[3.24 + .04 + .64 + 4.84 + .04]}{4}}$$

$$= \sqrt{\frac{8.8}{4}}$$

$$= \sqrt{2.2}$$

$$= 1.48$$

The square of the standard deviation, s, is referred to as the *variance*. Variance is not discussed in this book.

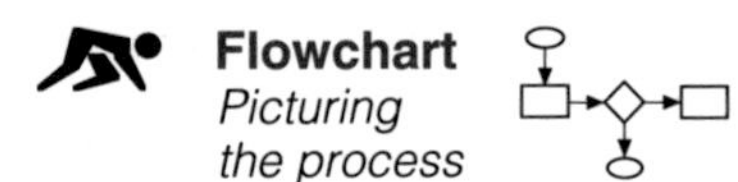

Why use it?

To allow a team to identify the actual flow or sequence of events in a process that any product or service follows. Flowcharts can be applied to anything from the travels of an invoice or the flow of materials, to the steps in making a sale or servicing a product.

What does it do?

- Shows unexpected complexity, problem areas, redundancy, unnecessary loops, and where simplification and standardization may be possible
- Compares and contrasts the actual versus the ideal flow of a process to identify improvement opportunities
- Allows a team to come to agreement on the steps of the process and to examine which activities may impact the process performance
- Identifies locations where additional data can be collected and investigated
- Serves as a training aid to understand the complete process

How do I do it?

1. **Determine the frame or boundaries of the process**
   - Clearly define where the process under study starts (input) and ends (final output).
   - Team members should agree to the level of detail they must show on the Flowchart to clearly understand the process and identify problem areas.

- The Flowchart can be a simple macro-flowchart showing only sufficient information to understand the general process flow or it might be detailed to show every finite action and decision point. The team might start out with a macro-flowchart and then add in detail later or only where it is needed.

2. **Determine the steps in the process**
   - Brainstorm a list of all major activities, inputs, outputs, and decisions on a flipchart sheet from the beginning of the process to the end.

3. **Sequence the steps**
   - Arrange the steps in the order they are carried out. Use Post-it™ notes so you can move them around. Don't draw in the arrows yet.

   Tip Unless you are flowcharting a new process, sequence what *is*, not what *should be* or the ideal. This may be difficult at first but is necessary to see where the probable causes of the problems are in the process.

4. **Draw the Flowchart using the appropriate symbols**

An oval is used to show the materials, information or action (inputs) to start the process or to show the results at the end (output) of the process.

A box or rectangle is used to show a task or activity performed in the process. Although multiple arrows may come into each box, usually only one output or arrow leaves each activity box.

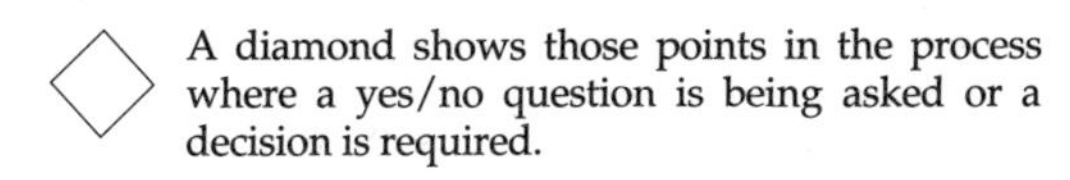

A diamond shows those points in the process where a yes/no question is being asked or a decision is required.

(A) A circle with either a letter or a number identifies a break in the Flowchart and is continued elsewhere on the same page or another page.

→ Arrows show the direction or flow of the process.

- Keep the Flowchart simple using the basic symbols listed above. As your experience grows, use other, more graphic symbols to represent the steps. Other symbols sometimes used include:
  - A half or torn sheet of paper for a report completed and/or filed
  - A can or computer tape wheel for data entry into a computer database
  - A large "D" or half circle to identify places in the process where there is a delay or wait for further action
- Be consistent in the level of detail shown.
  - A macro-level flowchart will show key action steps but no decision boxes.
  - An intermediate-level flowchart will show action and decision points.
  - A micro-level flowchart will show minute detail.
- Label each process step using words that are understandable to everyone.
- Add arrows to show the direction of the flow of steps in the process. Although not a rule, if you show all "yes" choices branching down and "no" choices branching to the left, it is easier to follow

the process. Preferences and space will later dictate direction.

- Don't forget to identify your work. Include the title of your process, the date the diagram was made, and the names of the team members.

5. **Test the Flowchart for completeness**
   - Are the symbols used correctly?
   - Are the process steps (inputs, outputs, actions, decisions, waits/delays) identified clearly?
   - Make sure every feedback loop is closed, i.e., every path takes you either back to or ahead to another step.
   - Check that every continuation point has a corresponding point elsewhere in the Flowchart or on another page of the Flowchart.
   - There is usually only one output arrow out of an activity box. If there is more than one arrow, you may need a decision diamond.
   - Validate the Flowchart with people who are not on the team and who carry out the process actions. Highlight additions or deletions they recommend. Bring these back to the team to discuss and incorporate into the final Flowchart.

6. **Finalize the Flowchart**
   - Is this process being run the way it should be?
   - Are people following the process as charted?
   - Are there obvious complexities or redundancies that can be reduced or eliminated?
   - How different is the current process from an ideal one? Draw an ideal Flowchart. Compare the two (current versus ideal) to identify discrepancies and opportunities for improvements.

Variations

The type of Flowchart just described is sometimes referred to as a "detailed" flowchart because it includes, in detail, the inputs, activities, decision points, and outputs of any process. Four other forms, described below, are also useful.

Macro Flowchart

Refer to the third bulleted item in Step 1 of this section for a description. For a graphic example, see Step 2 of the Improvement Storyboard in the Problem-Solving/ Process Improvement Model section.

Top-down Flowchart

This chart is a picture of the major steps in a work process. It minimizes the detail to focus only on those steps essential to the process. It usually does not include inspection, rework, and other steps that result in quality problems. Teams sometimes study the top-down flowchart to look for ways to simplify or reduce the number of steps to make the process more efficient and effective.

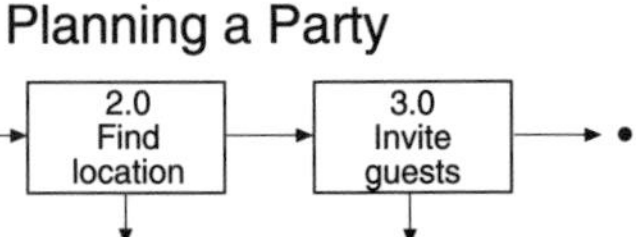

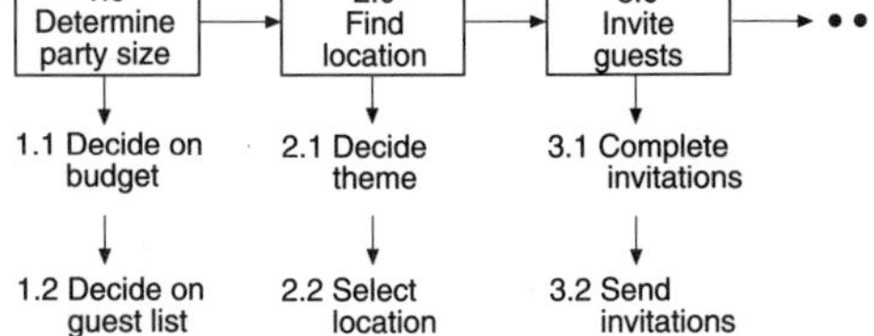

Deployment Flowchart

This chart shows the people or departments responsible and the flow of the process steps or tasks they are assigned. It is useful to clarify roles and track accountability as well as to indicate dependencies in the sequence of events.

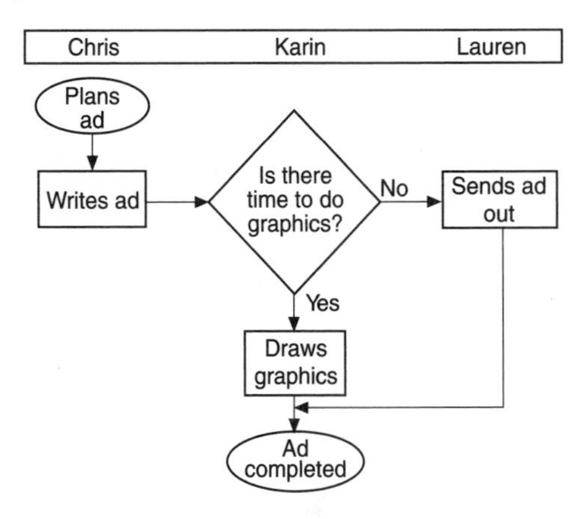

Workflow Flowchart

This type of chart is used to show the flow of people, materials, paperwork, etc., within a work setting. When redundancies, duplications, and unnecessary complexities are identified in a path, people can take action to reduce or eliminate these problems.

Flowchart

Proposed Patient Appointment Procedure

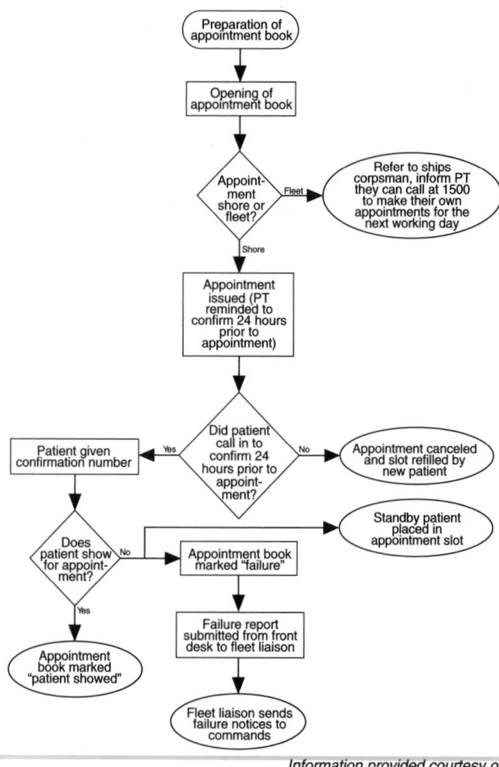

Information provided courtesy of U.S. Navy, Naval Dental Center, San Diego

Force Field Analysis

Positives & negatives of change

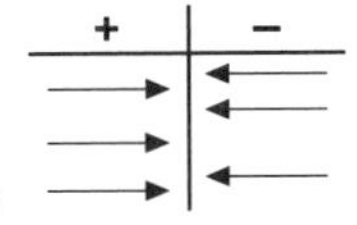

Why use it?

To identify the forces and factors in place that support or work against the solution of an issue or problem so that the positives can be reinforced and/or the negatives eliminated or reduced.

What does it do?

- Presents the "positives" and "negatives" of a situation so they are easily compared
- Forces people to think together about all the aspects of making the desired change a permanent one
- Encourages people to agree about the relative priority of factors on each side of the "balance sheet"
- Encourages honest reflection on the real underlying roots of a problem and its solution

How do I do it?

1. **Draw a large letter "T" on a flipchart**

   a) At the top of the T, write the issue or problem that you plan to analyze.

   - To the far right of the top of the T, write a description of the ideal situation you would like to achieve.

   b) Brainstorm the forces that are driving you towards the ideal situation. These forces may be internal or external. List them on the left side.

c) Brainstorm the forces that are restraining movement toward the ideal state. List them on the right side.

2. **Prioritize the driving forces that can be strengthened or identify restraining forces that would allow the most movement toward the ideal state if they were removed**

   - Achieve consensus through discussion or by using ranking methods such as Nominal Group Technique and Multivoting.

   Tip When choosing a target for change, remember that simply pushing the positive factors for a change can have the opposite effect. It is often more helpful to remove barriers. This tends to break the "change bottleneck" rather than just pushing on all the good reasons to change.

Force Field

Fear of Public Speaking

Ideal state: To speak confidently, clearly, and concisely in any situation.

+ Driving Forces	Restraining Forces –
Increases self-esteem →	← Past embarrassments
Helps career →	← Afraid to make mistakes
Communicates ideas →	← Lack of knowledge on the topic
Contributes to a plan/solution →	← Afraid people will be indifferent
Encourages others to speak →	← Afraid people will laugh
Helps others to change →	← May forget what to say
Increases energy of group →	← Too revealing of personal thoughts
Helps clarify speaker's ideas by getting feedback from others →	← Afraid of offending group
Hams can be hams (recognition from others) →	← Fear that nervousness will show
Helps others to see new perspective →	← Lack of confidence in personal appearance

Note: The Force Field can help individuals and teams select targets for change. Generally, when you focus on restraining forces, not driving forces, this works best. For example, using index cards for key points may reduce the fear "May forget what to say."

Histogram

Process centering, spread, and shape

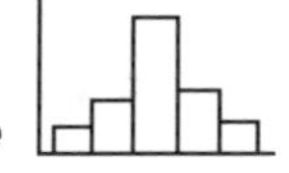

Why use it?

To summarize data from a process that has been collected over a period of time, and graphically present its frequency distribution in bar form.

What does it do?

- Displays large amounts of data that are difficult to interpret in tabular form
- Shows the relative frequency of occurrence of the various data values
- Reveals the centering, variation, and shape of the data
- Illustrates quickly the underlying distribution of the data
- Provides useful information for predicting future performance of the process
- Helps to indicate if there has been a change in the process
- Helps answer the question "Is the process capable of meeting my customer requirements?"

How do I do it?

1. **Decide on the process measure**
   - The data should be variable data, i.e., measured on a continuous scale. For example: temperature, time, dimensions, weight, speed.

2. **Gather data**
   - Collect at least 50 to 100 data points if you plan on looking for patterns and calculating the distribution's centering (mean), spread (variation), and shape. You might also consider collecting data for a specified period of time: hour, shift, day, week, etc.
   - Use historical data to find patterns or to use as a baseline measure of past performance.

3. **Prepare a frequency table from the data**

   a) Count the number of data points, n, in the sample.

9.9	9.3	10.2	9.4	10.1	9.6	9.9	10.1	9.8
9.8	9.8	10.1	9.9	9.7	9.8	9.9	10.0	9.6
9.7	9.4	9.6	10.0	9.8	9.9	10.1	10.4	10.0
10.2	10.1	9.8	10.1	10.3	10.0	10.2	9.8	10.7
9.9	10.7	9.3	10.3	9.9	9.8	10.3	9.5	9.9
9.3	10.2	9.2	9.9	9.7	9.9	9.8	9.5	9.4
9.0	9.5	9.7	9.7	9.8	9.8	9.3	9.6	9.7
10.0	9.7	9.4	9.8	9.4	9.6	10.0	10.3	9.8
9.5	9.7	10.6	9.5	10.1	10.0	9.8	10.1	9.6
9.6	9.4	10.1	9.5	10.1	10.2	9.8	9.5	9.3
10.3	9.6	9.7	9.7	10.1	9.8	9.7	10.0	10.0
9.5	9.5	9.8	9.9	9.2	10.0	10.0	9.7	9.7
9.9	10.4	9.3	9.6	10.2	9.7	9.7	9.7	10.7
9.9	10.2	9.8	9.3	9.6	9.5	9.6	10.7	

   In this example, there are 125 data points, n = 125.

   b) Determine the range, R, for the entire sample.

   The range is the smallest value in the set of data subtracted from the largest value. For our example:

$$R = X_{max} - X_{min} = 10.7 - 9.0 = 1.7$$

   c) Determine the number of class intervals, k, needed.

- Method 1: Take the square root of the total number of data points and round to the nearest whole number.

$$k = \sqrt{125} = 11.18 = 11 \text{ intervals}$$

- Method 2: Use the table below to provide a guideline for dividing your sample into a reasonable number of classes.

Number of Data Points	Number of Classes (k)
Under 50	5 – 7
50 – 100	6 – 10
100 – 250	7 – 12
Over 250	10 – 20

For our example, 125 data points would be divided into 7–12 class intervals.

Tip These two methods are general rules of thumb for determining class intervals. In both methods, consider using $k = 10$ class intervals for ease of "mental" calculation.

Tip The number of intervals can influence the pattern of the sample. Too few intervals will produce a tight, high pattern. Too many intervals will produce a spread out, flat pattern.

d) Determine the class width, H.

- The formula for this is:

$$H = \frac{R}{k} = \frac{1.7}{10} = .17$$

- Round your number to the nearest value with the same decimal numbers as the original sample. In our example, we would round up to .20. It is useful to have intervals defined to one more decimal place than the data collected.

e) Determine the class boundaries, or end points.

- Use the smallest individual measurement in the sample, or round to the next appropriate lowest round number. This will be the lower end point for the *first* class interval. In our example this would be 9.0.
- Add the class width, H, to the lower end point. This will be the lower end point for the *next* class interval. For our example:

$$9.0 + H = 9.0 + .20 = 9.20$$

Thus, the first class interval would be 9.00 and everything up to, *but not including* 9.20, that is, 9.00 through 9.19. The second class interval would begin at 9.20 and be everything up to, but not including 9.40.

Tip Each class interval must be *mutually exclusive,* that is, every data point will fit into *one, and only one* class interval.

- Consecutively add the class width to the lowest class boundary until the k class intervals and/or the range of all the numbers are obtained.

f) Construct the frequency table based on the values you computed in item "e."

A frequency table based on the data from our example is shown below.

Class #	Class Boundaries	Mid-Point	Frequency	Total
1	9.00-9.19	9.1	I	1
2	9.20-9.39	9.3	~~IIII~~ IIII	9
3	9.40-9.59	9.5	~~IIII~~ ~~IIII~~ ~~IIII~~ I	16
4	9.60-9.79	9.7	~~IIII~~ ~~IIII~~ ~~IIII~~ ~~IIII~~ ~~IIII~~ II	27
5	9.80-9.99	9.9	~~IIII~~ ~~IIII~~ ~~IIII~~ ~~IIII~~ ~~IIII~~ ~~IIII~~ I	31
6	10.00-10.19	10.1	~~IIII~~ ~~IIII~~ ~~IIII~~ ~~IIII~~ II	22
7	10.20-10.39	10.3	~~IIII~~ ~~IIII~~ II	12
8	10.40-10.59	10.5	II	2
9	10.60-10.79	10.7	~~IIII~~	5
10	10.80-10.99	10.9		0

4. **Draw a Histogram from the frequency table**

- On the vertical line, (y axis), draw the frequency (count) scale to cover class interval with the highest frequency count.
- On the horizontal line, (x axis), draw the scale related to the variable you are measuring.
- For each class interval, draw a bar with the height equal to the frequency tally of that class.

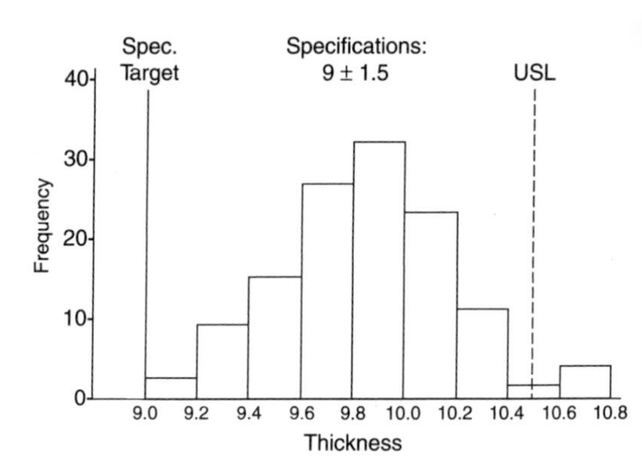

5. Interpret the Histogram

a) *Centering*. Where is the distribution centered? Is the process running too high? Too low?

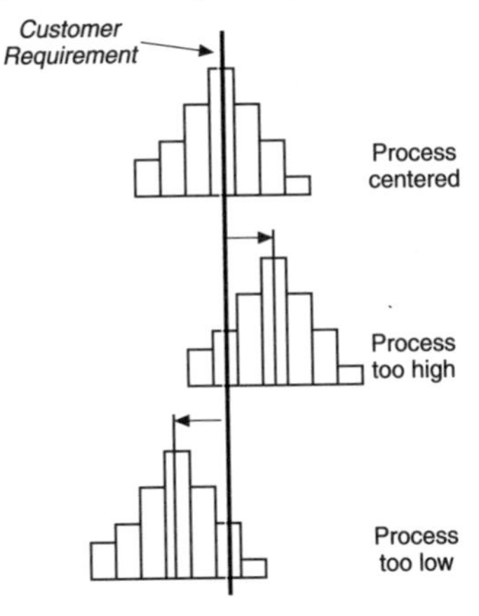

b) *Variation*. What is the variation or spread of the data? Is it too variable?

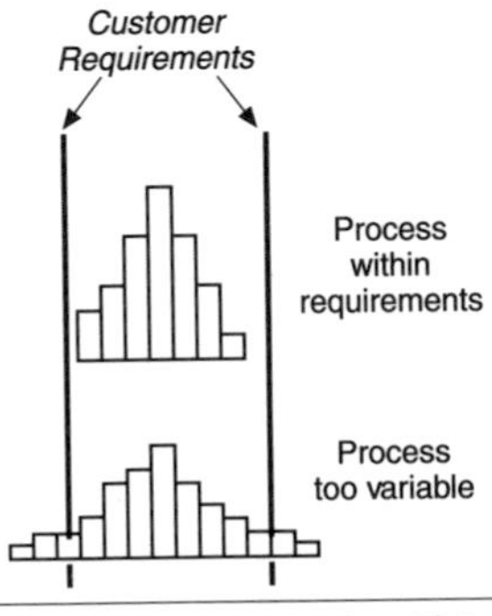

c) *Shape*. What is the shape? Does it look like a normal, bell-shaped distribution? Is it positively or negatively skewed, that is, more data values to the left or to the right? Are there twin (bi-modal) or multiple peaks?

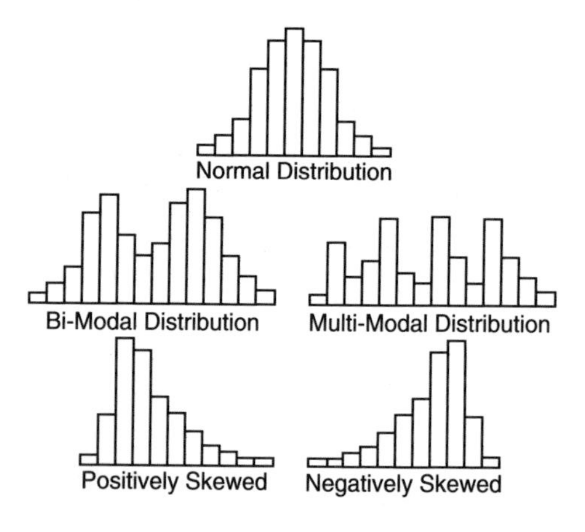

Tip Some processes are naturally skewed; don't expect every distribution to follow a bell-shaped curve.

Tip Always look for twin or multiple peaks indicating that the data is coming from two or more different sources, e.g., shifts, machines, people, suppliers. If this is evident, stratify the data.

d) *Process Capability*. Compare the results of your Histogram to your customer requirements or specifications. Is your process capable of meeting the requirements, i.e., is the Histogram centered on the target and within the specification limits?

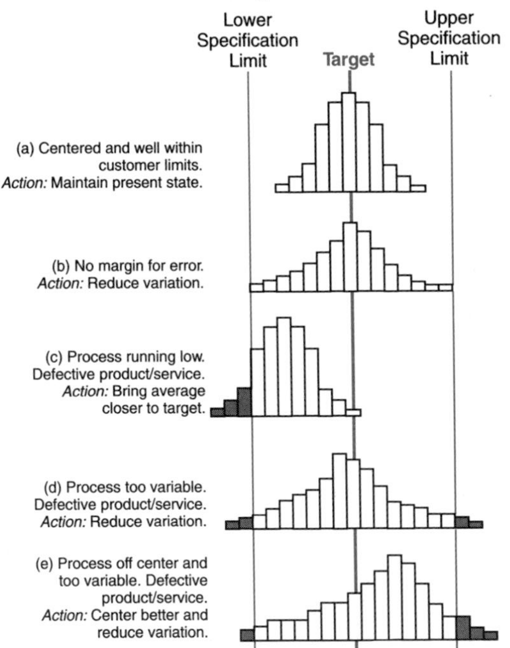

Tip Get suspicious of the accuracy of the data if the Histogram suddenly stops at one point (such as a specification limit) without some previous decline in the data. It could indicate that defective product is being sorted out and is not included in the sample.

Tip The Histogram is related to the Control Chart. Like a Control Chart, a normally distributed Histogram will have almost all its values within ±3 standard deviations of the mean. See Process Capability for an illustration of this.

Variations

Stem & Leaf Plot

This plot is a cross between a frequency distribution and Histogram. It exhibits the shape of a Histogram, but preserves the original data values—one of its key benefits! Data is easily recorded by writing the trailing digits in the appropriate row of leading digits.

Data as collected

Stem	Leaf
.05	7
.06	4
.07	5 7
.08	1 9 3 9
.09	7 4 8 2 6 9 4
.10	7 2 0 4 3 5 9
.11	3 1 9 3 7 3 8 6 6
.12	2 4 8 0 8 9 0 5
.13	2 5 2 7 7 6
.14	0 3 6 9
.15	4 7
.16	4

Data rank-ordered

Stem	Leaf
.05	7
.06	4
.07	5 7
.08	1 3 9 9
.09	2 4 4 6 7 8 9
.10	0 2 3 4 5 7 9
.11	1 3 3 3 6 6 7 8 9
.12	0 0 2 4 5 8 8 9
.13	2 2 5 6 7 7
.14	0 3 6 9
.15	4 7
.16	4

In this example, the smallest value is .057 and the largest value is .164. Using such a plot, it is easy to find the median and range of the data.

- Median = middle data value (or average of the two middle values) when the data is ranked from smallest to largest.

For this example, there are 52 data points. Therefore, the average of the 26th and 27th value will give the median value.

$$\text{Median} = (.113 + .116)/2 = .1145$$

- Range = Highest value – lowest value = .164 – .057 = .107

Histogram

Time Distribution of Calls

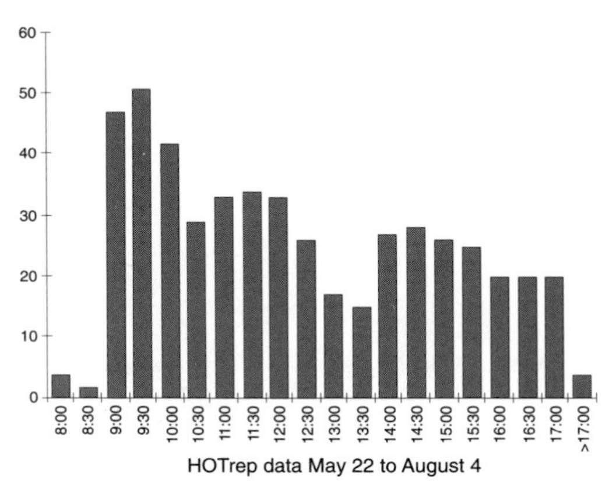

HOTrep data May 22 to August 4

Information provided courtesy of SmithKline Beecham

Note: The Histogram identified three peak calling periods at the beginning of the workday and before and after the traditional lunch hour. This can help the HOTreps synchronize staffing with their customer needs.

Interrelationship Digraph (ID)

Looking for drivers & outcomes

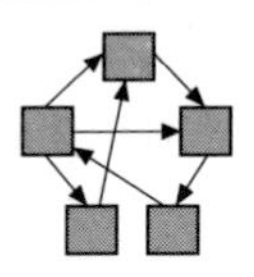

Why use it?

To allow a team to systematically identify, analyze, and classify the cause and effect relationships that exist among all critical issues so that key drivers or outcomes can become the heart of an effective solution.

What does it do?

- Encourages team members to think in multiple directions rather than linearly
- Explores the cause and effect relationships among all the issues, including the most controversial
- Allows the key issues to emerge naturally rather than allowing the issues to be forced by a dominant or powerful team member
- Systematically surfaces the basic assumptions and reasons for disagreements among team members
- Allows a team to identify root cause(s) even when credible data doesn't exist

How do I do it?

1. **Agree on the issue/problem statement**

> What are the issues related to reducing litter?

- If using an original statement, (it didn't come from a previous tool or discussion), create a complete sentence that is clearly understood and agreed on by team members.

- If using input from other tools, such as an Affinity Diagram, make sure that the goal under discussion is still the same and clearly understood.

2. **Assemble the right team**
   - The ID requires more intimate knowledge of the subject under discussion than is needed for the Affinity. This is important if the final cause and effect patterns are to be credible.
   - The ideal team size is generally 4–6 people. However, this number can be increased as long as the issues are still visible and the meeting is well facilitated to encourage participation and maintain focus.

3. **Lay out all of the ideas/issue cards that have either been brought from other tools or brainstormed**
   - Arrange 5–25 cards or notes in a large circular pattern, leaving as much space as possible for drawing arrows. Use large, bold printing, including a large number or letter on each idea for quick reference later in the process.

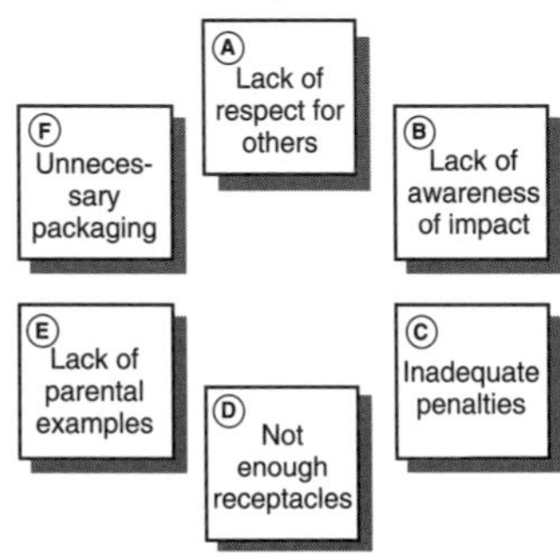

4. **Look for cause/influence relationships between all of the ideas and draw relationship arrows**

- Choose any of the ideas as a starting point. If all of the ideas are numbered or lettered, work through them in sequence.
- An outgoing arrow from an idea indicates that it is the stronger cause or influence.

Ask of each combination:

1) Is there a cause/influence relationship?
2) If yes, which direction of cause/influence is stronger?

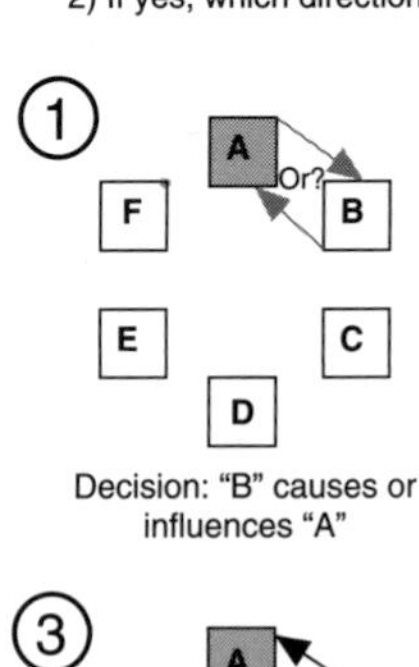

Decision: "B" causes or influences "A"

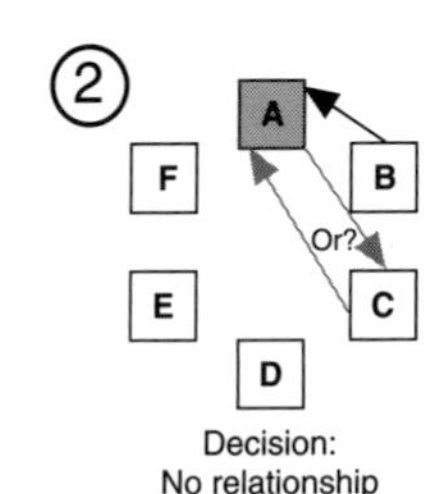

Decision: No relationship

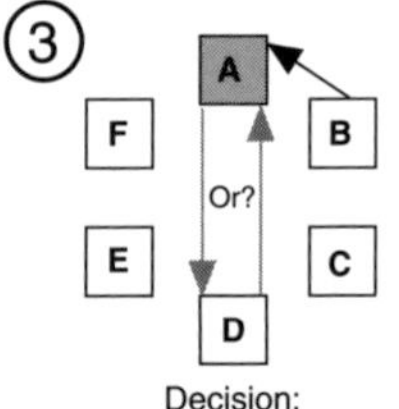

Decision: No relationship

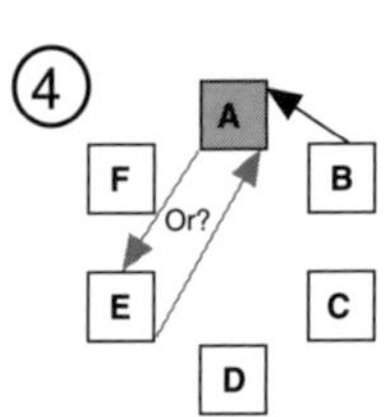

Decision: "E" causes or influences "A"

Continued next page

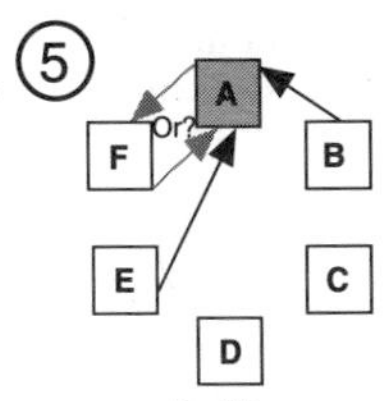

Decision:
No relationship.
"A" is completed.

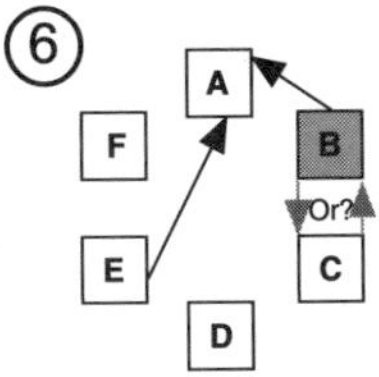

Decision: "B" causes or influences "C."
Now begin with "B" and repeat the questions for all remaining combinations.

Tip Draw only one-way relationship arrows in the direction of the stronger cause or influence. Make a decision on the stronger direction. ***Do not draw two-headed arrows.***

5. **Review and revise the first round ID**
   - Get additional input from people who are not on the team to confirm or modify the team's work. Either bring the paper version to others or reproduce it using available software. Use a different size print or a color marker to make additions or deletions.

6. **Tally the number of outgoing and incoming arrows and select key items for further planning**
   - Record and clearly mark next to each issue the number of arrows going in and out of it.
   - Find the item(s) with the highest number of *outgoing arrows* and the item(s) with the highest number of *incoming arrows*.
   - *Outgoing arrows.* A high number of outgoing arrows indicates an item that is a root cause or driver. This is *generally* the issue that teams tackle first.

- *Incoming arrows*. A high number of incoming arrows indicates an item that is a key outcome. This can become a focus for planning either as a meaningful measure of overall success or as a redefinition of the original issue under discussion.

Tip Use common sense when you select the most critical issues to focus on. Issues with very close tallies must be reviewed carefully but in the end, it is a judgment call, not science.

7. **Draw the final ID**
   - Identify visually both the *key drivers* (greatest number of outgoing arrows) and the *key outcomes* (greatest number of incoming arrows). Typical methods are double boxes or bold boxes.

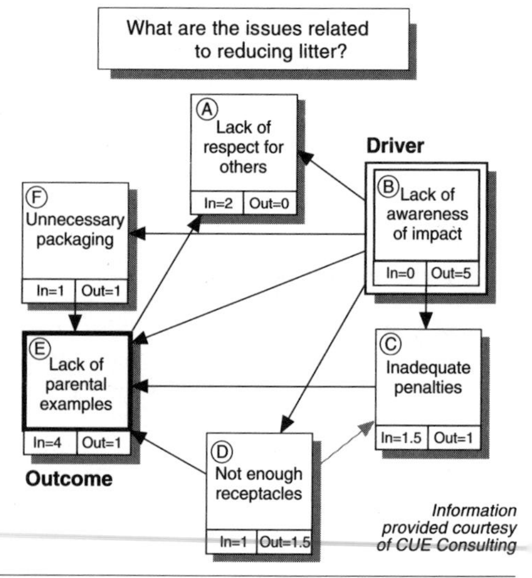

Information provided courtesy of CUE Consulting

Variations

When it is necessary to create a more orderly display of all of the relationships, a matrix format is very effective. The vertical (up) arrow is a driving cause and the horizontal (side) arrow is an effect. The example below has added symbols indicating the strength of the relationships.

The "total" column is the sum of all of the "relationship strengths" in each row. This shows that you are working on those items that have the strongest effect on the greatest number of issues.

ID – Matrix Format

	Logistic Support	Customer Satisfaction	Education & Training	Personnel Incentives	Leadership	Cause/ Driver ↑	Result/ Rider ←	Total
Logistic Support		⊙ ↑	○ ↑	△ ↑	○ ←	3	1	16
Customer Satisfaction	⊙ ←		○ ←	⊙ ←	○ ←	0	4	24
Education & Training	○ ←	○ ↑		○ ↑	⊙ ←	2	2	18
Personnel Incentives	△ ←	⊙ ↑	○ ←		⊙ ←	1	3	22
Leadership	○ ↑	○ ↑	⊙ ↑	⊙ ↑		4	0	24

Relationship Strength

⊙ = 9 Significant
○ = 3 Medium
△ = 1 Weak

Information provided courtesy of U.S. Air Force, Air Combat Command

Interrelationship Digraph

Issues Surrounding Implementation of the Business Plan

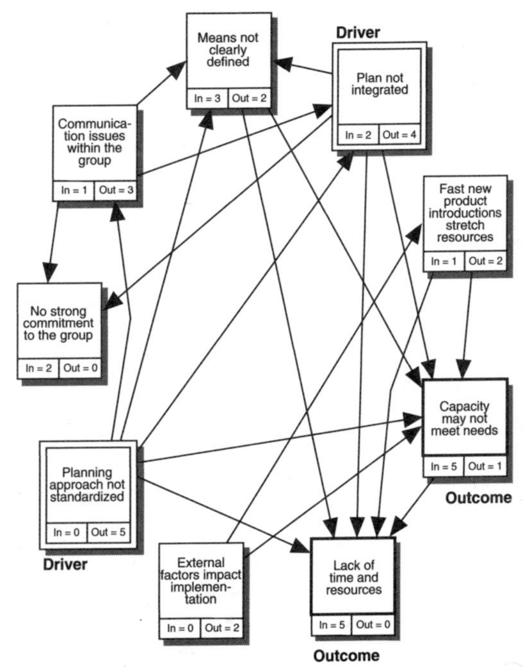

Information provided courtesy of Goodyear

Note: "The drivers" from this ID will be used as the goal in the Tree example shown at the end of the Tree Diagram/PDPC section.

Interrelationship Digraph

A Vision of Andover in the 21st Century

See next page for close up

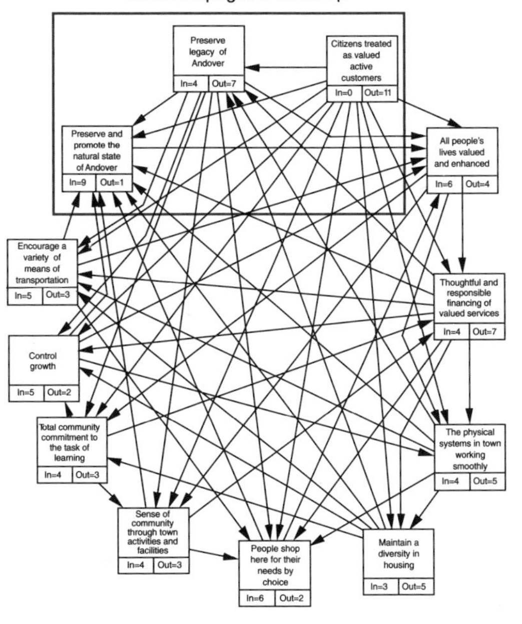

Information provided courtesy of Town of Andover, MA

Interrelationship Digraph

A Vision of Andover in the 21st Century

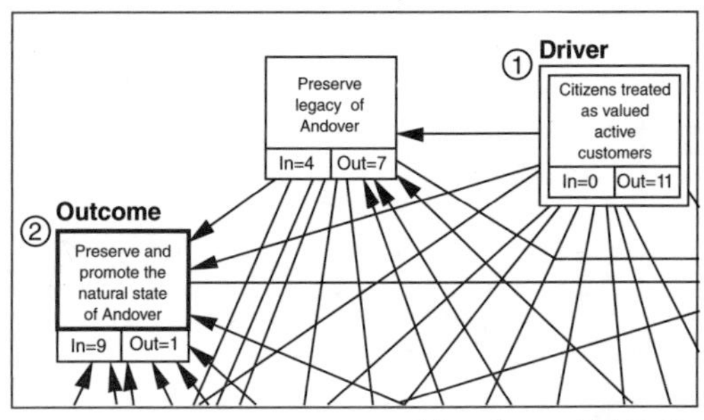

Information provided courtesy of Town of Andover, MA

① This is the driver. If the focus on the citizen as a customer becomes the core of the town's vision then everything else will be advanced.

② This is the primary outcome. It puts the preservation of nature in the town as a key indicator of the vision working.

Matrix Diagram

Finding relationships

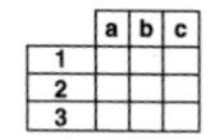

Why use it?

To allow a team or individual to systematically identify, analyze, and rate the presence and strength of relationships between two or more sets of information.

What does it do?

- Makes patterns of responsibilities visible and clear so that there is an even and appropriate distribution of tasks
- Helps a team get consensus on small decisions, enhancing the quality and support for the final decision
- Improves a team's discipline in systematically taking a hard look at a large number of important decision factors

Types of Matrices

Most Common

- *L-shaped matrix.* Two sets of items directly compared to each other or a single set compared to itself.

Orienting New Employees

Tasks / Resources	Tour facility	Review personnel & safety policies	Review business values	Introduce to team members
Human resources		○	△	
Division manager			⊙	
Supervisor	△	⊙	○	○
Associates	⊙	△	△	⊙

⊙ Primary responsibility
○ Team members
△ Resources

Conclusion: Supervisors and associates have taken on the orientation role rather than the traditional human resource function.

- *T-shaped matrix.* Two sets of items compared to a common third set.

Orienting New Employees

Communicate organization spirit	⦿	○	⦿	⦿
Communicate purpose of organization			⦿	⦿
Resolve practical concerns	⦿	⦿	○	⦿
Reduce anxiety	⦿	⦿	⦿	⦿
Goals / Resources — Tasks	Tour facility	Review personnel & safety policies	Review business values	Introduce to team members
Human resources		○	△	
Division manager			⦿	
Supervisor	△	⦿	○	○
Associates	⦿	△	△	⦿

Responsibility
⦿ Primary
○ Team members
△ Resources

Impact
⦿ High
○ Medium
△ Low

Conclusion: The most important purpose of orientation is to reduce anxiety, and the most effective tasks focus on the personal issues.

Uncommon

- *Y-shaped matrix.* Three sets of items compared to each other. It "bends" a T-shaped matrix to allow comparisons between items that are on the vertical axes.

Rarely Used

- *X-shaped matrix.* Four sets of items compared to each other. It is essentially two T-shaped matrices placed back to back.

- *C-shaped matrix*. Shows the intersection of three sets of data simultaneously. It is a three-dimensional graphic.
- You can find more complete information on the Y-, X-, and C-shaped matrix in *The Memory Jogger Plus+®*.

How do I do it?

1. Select the key factors affecting successful implementation

- The most important step is to choose the issues or factors to be compared. The format is secondary. Begin with the right issues and the best format will define itself. The most common use is the distribution of responsibilities within an L-shaped or T-shaped matrix.

2. Assemble the right team

- Select individuals that have the influence/power to realistically assess the chosen factors.

Tip When distributing responsibilities, include those people who will likely be involved in the assigned tasks or who can at least be part of a review team to confirm small group results.

3. Select an appropriate matrix format

- Base your choice of format on the number of sets of items and types of comparisons you need to make.

4. Choose and define relationship symbols

- The most common symbols in matrix analysis are ⊙, ○, △. Generally they are used to indicate:

⊙	=	High	=	9
○	=	Medium	=	3
△	=	Low	=	1

- The possible meanings of the symbols are almost endless. The only requirement is that the team comes to a clear understanding and creates an equally clear legend with the matrix.

5. **Complete the matrix**
   - If distributing responsibilities, use only one "primary responsibility" symbol to show ultimate accountability. All other core team members can be given secondary responsibilities.

   Tip Focus the quality of the decision in each matrix cell. Do not try to "stack the deck" by consciously building a pattern of decisions. Let these patterns emerge naturally.

   Tip Interpret the matrix using total numerical values only when it adds value. Often the visual pattern is sufficient to interpret the overall results.

Variations

The matrix is one of the most versatile tools available. The important skill to master is "matrix thinking." This approach allows a team to focus its discussion on related factors that are explored thoroughly. The separate conclusions are then brought together to create high-quality decisions. Use your creativity in determining which factors affect each other, and in choosing the matrix format that will help focus the discussion toward the ultimate decision.

Matrix

Logistics Annual Plan

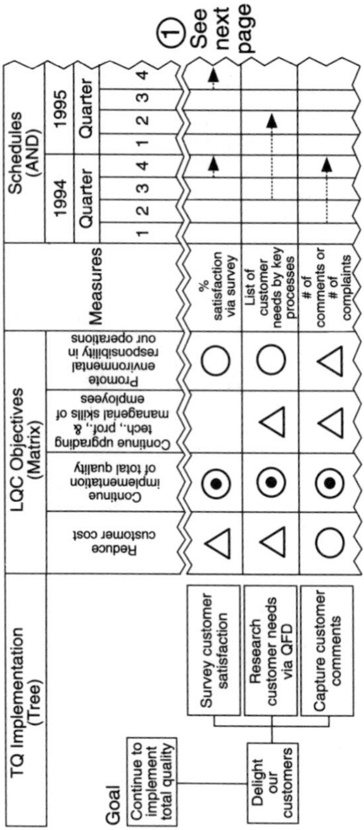

TQ Implementation (Tree)			LQC Objectives (Matrix)				Measures	Schedules (AND)							
								1994				1995			
								Quarter				Quarter			
Goal			Reduce customer cost	Continue implementation of total quality	Continue upgrading tech., prof., & managerial skills of employees	Promote environmental responsibility in our operations		1	2	3	4	1	2	3	4
Continue to implement total quality	Delight our customers	Survey customer satisfaction	△	◉		○	% satisfaction via survey				▶				▶
		Research customer needs via QFD	△	◉	△	○	List of customer needs by key processes			·····	·····	·····	▶		
		Capture customer comments	○	◉	△	△	# of comments or # of complaints		·····	·····	▶				

① See next page

◉ = 9 Strong influence/relationship
○ = 3 Some influence/relationship
△ = 1 Weak influence/relationship
Blank = No influence/relationship

Information provided courtesy of Bell Canada

Matrix

Logistics Annual Plan

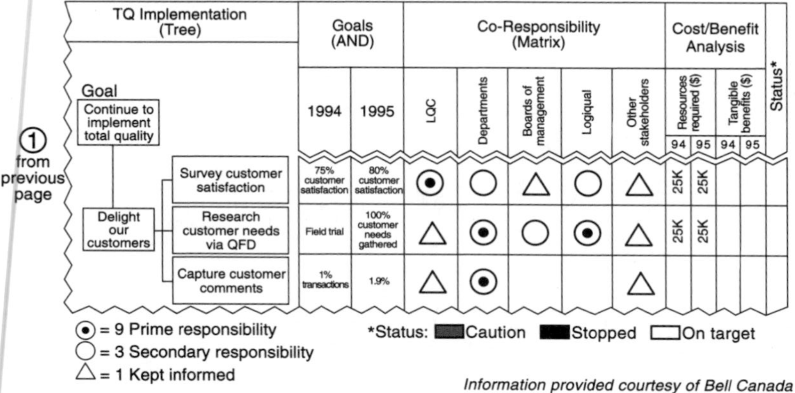

① from previous page

TQ Implementation (Tree)	Goals (AND)		Co-Responsibility (Matrix)					Cost/Benefit Analysis				Status*
	1994	1995	LQC	Departments	Boards of management	Logiqual	Other stakeholders	Resources required ($) 94	Resources required ($) 95	Tangible benefits ($) 94	Tangible benefits ($) 95	
Goal: Continue to implement total quality → Delight our customers → Survey customer satisfaction	75% customer satisfaction	80% customer satisfaction	◉	○	△	○	△	25K	25K			
Research customer needs via QFD	Field trial	100% customer needs gathered	△	◉	○	◉	△	25K	25K			
Capture customer comments	1% transactions	1.9%	△	◉			△					

◉ = 9 Prime responsibility
○ = 3 Secondary responsibility
△ = 1 Kept informed

*Status: Caution Stopped On target

Information provided courtesy of Bell Canada

Nominal Group Technique (NGT)

Ranking for consensus

A. 2+2
B. 1+1
C. 3+4
D. 4+3

Why use it?

Allows a team to quickly come to a consensus on the relative importance of issues, problems, or solutions by completing individual importance rankings into a team's final priorities.

What does it do?

- Builds commitment to the team's choice through equal participation in the process
- Allows every team member to rank issues without being pressured by others
- Puts quiet team members on an equal footing with more dominant members
- Makes a team's consensus (or lack of it) visible; the major causes of disagreement can be discussed

How do I do it?

1. **Generate the list of issues, problems, or solutions to be prioritized**
   - In a new team with members who are not accustomed to team participation, it may feel safer to do written, silent brainstorming, especially when dealing with sensitive topics.

2. **Write statements on a flipchart or board**

3. **Eliminate duplicates and/or clarify meanings of any of the statements**
   - As a leader, *always* ask for the team's permission and guidance when changing statements.

4. **Record the final list of statements on a flipchart or board**

   Example: Why does the department have inconsistent output?

A	Lack of training
B	No documented process
C	Unclear quality standards
D	Lack of cooperation with other departments
E	High turnover

   - Use letters rather than numbers to identify each statement so that team members do not get confused by the ranking process that follows.

5. **Each team member records the corresponding letters on a piece of paper and rank orders the statements**

   Example: Larry's sheet of paper looks like this:

A	4
B	5
C	3
D	1
E	2

   - This example uses "5" as the most important ranking and "1" as the least important. Since individual rankings will later be combined, this "reverse order" minimizes the effect of team members leaving some statements blank. Therefore, a blank (value = 0) would not, in effect, increase its importance.

6. Combine the rankings of all team members

	Larry	Nina	Norm	Paige	Si		Total
A	4	5	2	2	1	=	14
B	5	4	5	3	5	=	22
C	3	1	3	4	4	=	15
D	1	2	1	5	2	=	11
E	2	3	4	1	3	=	13

"No documented process," B, would be the highest priority. The team would work on this first and then move through the rest of the list as needed.

Variations

One Half Plus One

When dealing with a large number of choices it may be necessary to limit the number of items ranked. The "one half plus one" approach would rank only a portion of the total. For example, if 20 ideas were generated, then team members would rank only the top 11 choices. If needed, this process could be repeated with the remaining 9 items, ranking the top 5 or 6 items, (half of 9 = 4.5 + 1 = 5.5), until a manageable number are identified.

Weighted Multivoting

Each team member *rates, not ranks,* the relative importance of choices by distributing a value, e.g., 100 points, across the options. Each team member can distribute this value among as many or as few choices as desired.

Example:

	Larry	Nina	Norm	Paige	Si		Total
A	20		10			=	30
B	40	80	50	100	45	=	315
C	30	5	10		25	=	70
D		5	10		20	=	35
E	10	10	20		10	=	50

With large numbers of choices, or when the voting for the top choices is very close, this process can be repeated for an agreed upon number of items. Stop when the choice is clear.

Pareto Chart

Focus on key problems

Why use it?

To focus efforts on the problems that offer the greatest potential for improvement by showing their relative frequency or size in a descending bar graph.

What does it do?

- Helps a team to focus on those causes that will have the greatest impact if solved
- Based on the proven Pareto principle: 20% of the sources cause 80% of any problem
- Displays the relative importance of problems in a simple, quickly interpreted, visual format
- Helps prevent "shifting the problem" where the "solution" removes some causes but worsens others
- Progress is measured in a highly visible format that provides incentive to push on for more improvement

How do I do it?

1. **Decide which problem you want to know more about**

   Example: Consider the case of HOTrep, an internal computer network help line: Why do people call the HOTrep help line; what problems are people having?

2. **Choose the causes or problems that will be monitored, compared, and rank ordered by brainstorming or with existing data**

   a) Brainstorming

Example: What are typical problems that users ask about on the HOTrep help line?

b) Based on existing data

Example: What problems in the last month have users called in to the HOTrep help line?

3. **Choose the most meaningful unit of measurement such as frequency or cost**
   - Sometimes you don't know before the study which unit of measurement is best. Be prepared to do both frequency and cost.

   Example: For the HOTrep data the most important measure is frequency because the project team can use the information to simplify software, improve documentation or training, or solve bigger system problems.

4. **Choose the time period for the study**
   - Choose a time period that is long enough to represent the situation. Longer studies don't always translate to *better* information. Look first at volume and variety within the data.
   - Make sure the scheduled time is typical in order to take into account seasonality or even different patterns within a given day or week.

   Example: Review HOTrep help line calls for 10 weeks (May 22–August 4).

5. **Gather the necessary data on each problem category either by "real time" or reviewing historical data**
   - Whether data is gathered in "real time" or historically, check sheets are the easiest method for collecting data.

   Example: Gathered HOTrep help line calls data based on the review of incident reports (historical).

Tip *Always* include with the source data and the final chart, the identifiers that indicate the source, location, and time period covered.

6. **Compare the relative frequency or cost of each problem category**

   Example:

Problem Category	Frequency	Percent (%)
Bad configuration	3	1
Boot problems	68	33
File problems	8	4
Lat. connection	20	10
Print problems	16	8
Reflection hang	24	12
Reflection sys. integrity	11	5
Reflections misc.	6	3
System configuration	16	8
System integrity	19	9
Others	15	7
Total	**206**	

7. **List the problem categories on the horizontal line and frequencies on the vertical line**
   - List the categories in descending order from left to right on the horizontal line with bars above each problem category to indicate its frequency or cost. List the unit of measure on the vertical line.

8. **(Optional) Draw the cumulative percentage line showing the portion of the total that each problem category represents**

   a) On the vertical line, (opposite the raw data, #, $, etc.), record 100% opposite the total number and 50% at the halfway point. Fill in the remaining percentages drawn to scale.

b) Starting with the highest problem category, draw a dot or mark an x at the upper righthand corner of the bar.

- Add the total of the next problem category to the first and draw a dot above that bar showing both the cumulative number and percentage. Connect the dots and record the remaining cumulative totals until 100% is reached.

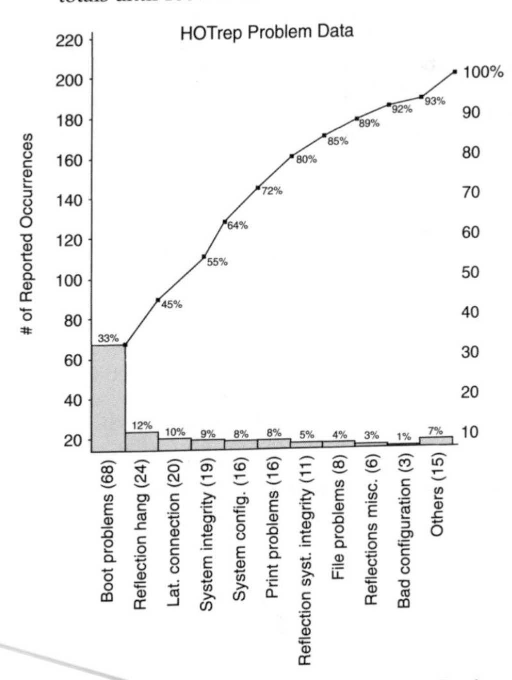

Information provided courtesy of SmithKline Beecham

9. **Interpret the results**
   - *Generally*, the tallest bars indicate the biggest contributors to the overall problem. Dealing with these problem categories first therefore makes common sense. *But*, the most frequent or expensive is not always the most important. Always ask: What has the most impact on the goals of our business and customers?

Variations

The Pareto Chart is one of the most widely and creatively used improvement tools. The variations used most frequently are:

A. **Major Cause Breakdowns** in which the "tallest bar" is broken into subcauses in a second, linked Pareto.

B. **Before and After** in which the "new Pareto" bars are drawn side by side with the original Pareto, showing the effect of a change. It can be drawn as one chart or two separate charts.

C. **Change the Source of Data** in which data is collected on the same problem but from different departments, locations, equipment, and so on, and shown in side-by-side Pareto Charts.

D. **Change Measurement Scale** in which the same categories are used but measured differently. Typically "cost" and "frequency" are alternated.

Pareto

A. Major Cause Breakdowns

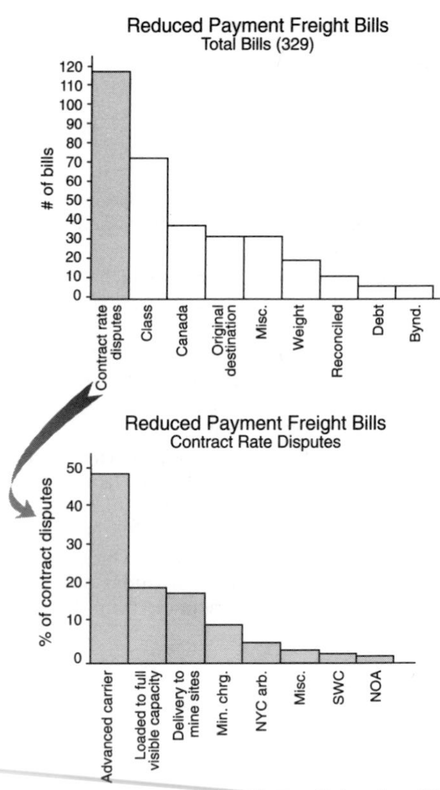

Information provided courtesy of Goodyear

Pareto

B. Before and After

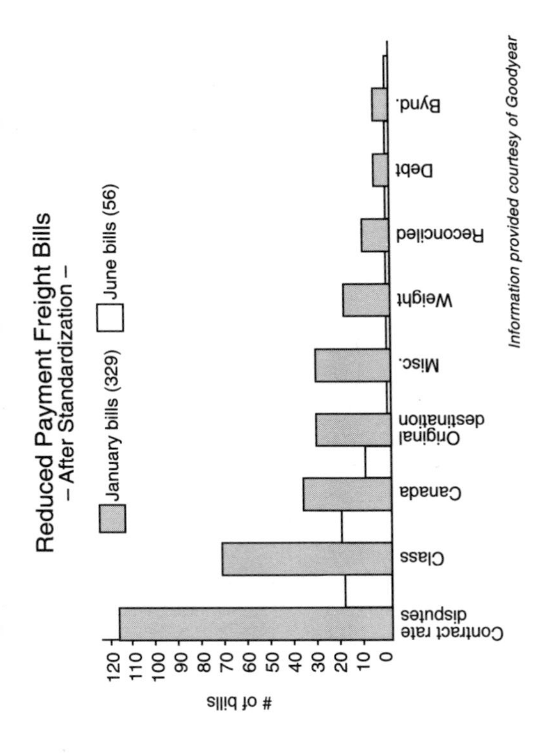

C. Change the Source of Data

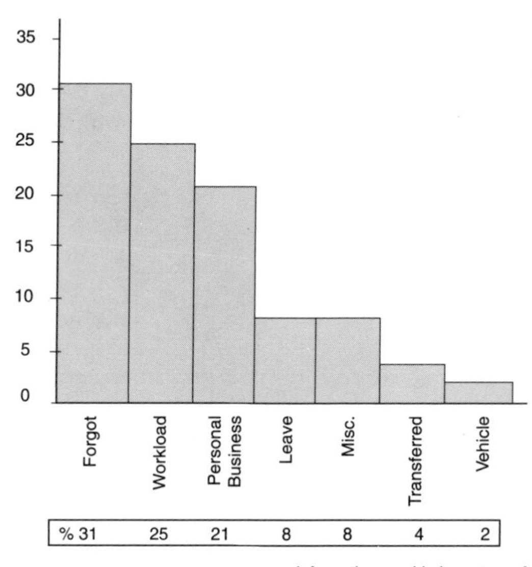

% 31	25	21	8	8	4	2

Information provided courtesy of U.S. Navy, Naval Dental Center, San Diego

C. Change the Source of Data

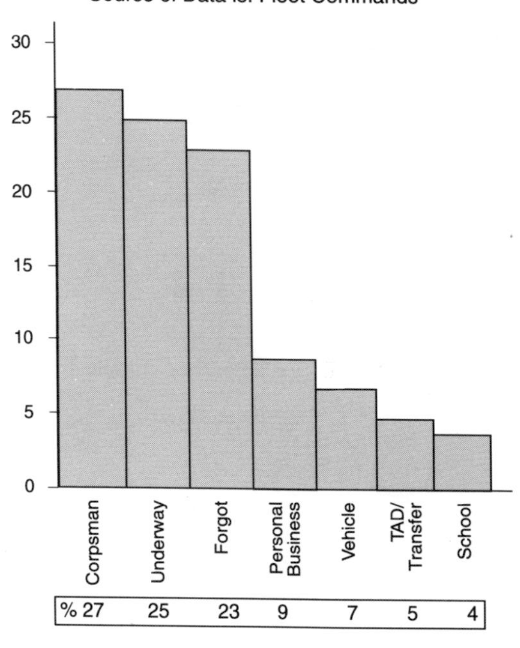

% 27	25	23	9	7	5	4

Information provided courtesy of U.S. Navy, Naval Dental Center, San Diego

Pareto

D. Change Measurement Scale

Field Service Customer Complaints

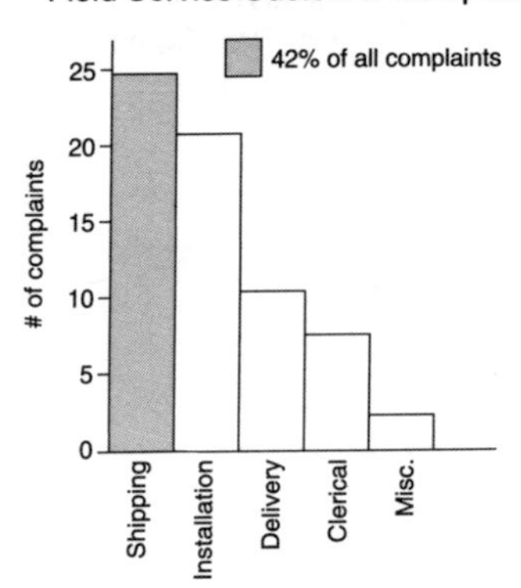

Cost to Rectify Field Service Complaints

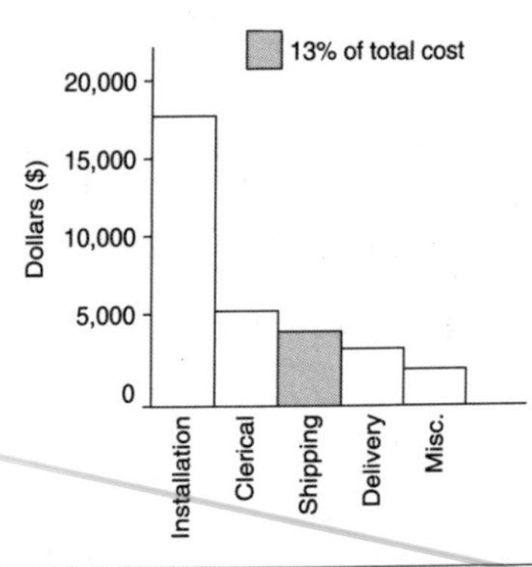

Prioritization Matrices

Weighing your options

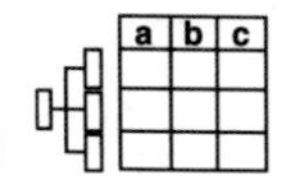

Why use it?

To narrow down options through a systematic approach of comparing choices by selecting, weighting, and applying criteria.

What does it do?

- Quickly surfaces basic disagreements so they may be resolved up front
- Forces a team to focus on the best thing(s) to do, and not everything they could do, dramatically increasing the chances for implementation success
- Limits "hidden agendas" by surfacing the criteria as a necessary part of the process
- Increases the chance of follow-through because consensus is sought at each step in the process (from criteria to conclusions)
- Reduces the chances of selecting someone's "pet project"

How do I do it?

There are three methods for constructing Prioritization Matrices. The outline that follows indicates typical situations for using each method. Only the "Full Analytical Criteria Method" is discussed here. The others are covered fully in *The Memory Jogger Plus+®*.

Full Analytical Criteria Method

Typically use when:

- Smaller teams are involved (3–8 people)
- Options are few (5–10 choices)
- There are relatively few criteria (3–6 items)
- Complete consensus is needed
- The stakes are high if the plan fails

Consensus Criteria Method

This method follows the same steps as in the Full Analytical Criteria Method except the Consensus Criteria Method uses a combination of weighted voting, and ranking is used instead of paired comparisons.

Typically use when:

- Larger teams are involved (8 or more people)
- Options are many (10–20 choices)
- There are a significant number of criteria (6–15 items)
- Quick consensus is needed to proceed

Combination ID/Matrix Method

This method is different from the other two methods because it is based on cause and effect, rather than criteria.

Typically use when:

- Interrelationships among options are high and finding the option with the greatest impact is critical

Full Analytical Criteria Method

1. **Agree on the ultimate goal to be achieved in a clear, concise sentence**
   - If no other tools are used as input, produce a clear goal statement through consensus. This statement strongly affects which criteria are used.

   > Choose the most enjoyable vacation for the whole family

2. **Create the list of criteria**
   - Brainstorm the list of criteria or review previous documents or guidelines that are available, e.g., corporate goals, budget-related guidelines.

   > - Cost
   > - Educational value
   > - Diverse activity
   > - Escape reality

   Tip The team *must reach consensus* on the final criteria and their meanings or the process is likely to fail!

3. **Using an L-shaped matrix, weight each criterion against each other**
   - Reading across from the vertical axis, compare each criterion to those on the horizontal axis.
   - Each time a weight (e.g., 1, 5, 10) is recorded in a row cell, its reciprocal value (e.g., $1/5$, $1/10$) must be recorded in the corresponding column cell.
   - Total each horizontal row and convert to a relative decimal value known as the "criteria weighting."

Criterion vs. Criterion

Criteria \ Criteria	Cost	Educ. value	Diverse activity	Escape reality	Row Total	Relative Decimal Value
Cost		1/5	1/10	5	5.3	.15
Educ. value	5		1/5	5	10.2	.28
Diverse activity	10	5		5	20	.55
Escape reality	1/5	1/5	1/5		.60	.02
				Grand Total	36.1	

1 = Equally important
5 = More important
10 = Much more important
1/5 = Less Important
1/10 = Much less important

Row Total
Rating scores added
Grand Total
Row totals added
Relative Decimal Value
Each row total ÷ by the grand total

4. **Compare ALL options relative to each weighted criterion**
   - *For each criterion,* create an L-shaped matrix with all of the options on both the vertical and horizontal axis and the criteria listed in the lefthand corner of the matrix. ***There will be as many options matrices as there are criteria to be applied.***
   - Use the same rating scale (1, 5, 10) as in Step 3, *BUT* customize the wording for each criterion.
   - The relative decimal value is the "option rating."

Options vs. Each Criterion (Cost Criterion)

Cost	Disney World	Gettys-burg	New York City	Uncle Henry's	Row Total	Relative Decimal Value
Disney World		1/5	5	1/10	5.3	.12
Gettys-burg	5		10	1/5	15.2	.33
New York City	1/5	1/10		1/10	.40	.01
Uncle Henry's	10	5	10		25	.54
				Grand Total	45.9	

1 = Equal cost
5 = Less expensive
10 = Much less expensive
1/5 = More expensive
1/10 = Much more expensive

Continue Step 4 through three more Options/Criterion matrices, like this:

Escape reality

Crt.	Options			
Options				

Diverse activity

Crt.	Options			
Options				

Educational value

Crt.	Options			
Options				

Tip The whole number (1, 5, 10) must always represent a desirable rating. In some cases this may mean "less," e.g., cost, in others this may mean "more," e.g., tasty.

5. **Using an L-shaped summary matrix, compare each option based on all criteria combined**

- List all criteria on the horizontal axis and all options on the vertical axis.
- In each matrix cell multiply the "criteria weighting" of each criterion (decimal value from Step 3) by the "option rating" (decimal value from Step 4). This creates an "option score."
- Add each option score across all criteria for a row total. Divide each row total by the grand total and convert to the final decimal value. Compare these decimal values to help you decide which option to pursue.

Summary Matrix
Options vs. All Criteria

Criteria / Optns.	Cost (.15)	Educational value (.28)	Diverse activity (.55)	Escape reality (.02)	Row Total	Relative Decimal Value (RT ÷ GT)
Disney World	.12 x .15 (.02)	.24 x .28 (.07)	.40 x .55 (.22)	.65 x .02 (.01)	.32	.32
Gettysburg	.33 x .15 (.05)	.37 x .28 (.10)	.10 x .55 (.06)	.22 x .02 (0)	.22	.22
New York City	.01 x .15 (0)	.37 x .28 (.10)	.49 x .55 (.27)	.12 x .02 (0)	.37	.38
Uncle Henry's	.54 x .15 (.08)	.01 x .28 (0)	.01 x .55 (.01)	.01 x .02 (0)	.09	.09
				Grand Total	1.00	

.54 (from Step 4 matrix) x .15 (from Step 3 matrix)

(.08)
Option score

6. **Choose the best option(s) across all criteria**

Tip While this is more systematic than traditional decision making, it is not a science. Use common sense and judgment when options are rated very closely, but be open to non-traditional conclusions.

Variations

See *The Memory Jogger Plus+®* for full explanations of both the Consensus Criteria Method and the Combination ID/Matrix Method. The Full Analytical Criteria Method, illustrated in this book, is recommended because it encourages full discussion and consensus on critical issues. The Full Analytical Criteria Method is a simplified adaptation of an even more rigorous model known as the Analytical Hierarchy Process. It is based on the work of Thomas Saaty, which he describes in his book *Decision Making for Leaders*. In any case, use common sense to know when a situation is important enough to warrant such thorough processes.

Prioritization

Choosing a Standard Corporate Spreadsheet Program

① Weighting criteria (described in Step 3)

This is a portion of a full matrix with 14 criteria in total.

Criteria	Best use of hardware	Ease of use	Maximum functionality	Best performance	Total (14 criteria)	Relative Decimal Value
Best use of hardware		.20	.10	.20	3.7	.01
Ease of use	5.0		.20	.20	35.4	.08
Maximum functionality	10.0	5.0		5.0	69.0	.17
Best performance	5.0	5.0	.20		45.2	.11
				Grand Total (14 criteria)	418.1	

Information provided courtesy of Novacor Chemicals

Note: This constructed example, illustrated on three pages, represents only a portion of the prioritization process and only a portion of Novacor's spreadsheet evaluation process. Novacor Chemicals assembled a 16-person team, comprised mainly of system users and some information systems staff. The team developed and weighted 14 standard criteria and then applied them to choices in word processing, spreadsheet, and presentation graphics programs.

This example continued next page

Prioritization

Choosing a Standard Corporate Spreadsheet Program (cont.)

② Comparing options (described in Step 4)

These are just 2 of 14 matrices.

Best integration –internal	Program A	Program B	Program C	Total	Relative Decimal Value
Program A		1.00	1.00	2.00	.33
Program B	1.00		1.00	2.00	.33
Program C	1.00	1.00		2.00	.33
			Grand Total	6.00	

Lowest ongoing cost	Program A	Program B	Program C	Total	Relative Decimal Value
Program A		.10	.20	.30	.02
Program B	10.00		5.00	15.00	.73
Program C	5.00	.20		5.20	.25
			Grand Total	20.50	

Information provided courtesy of Novacor Chemicals

This example continued next page

Prioritization

Choosing a Standard Corporate Spreadsheet Program (cont.)

③ Summarize Option Ratings Across All Criteria (described in Step 5)

This is a portion of a full matrix with 14 criteria in total.

Criteria / Options	Easy to use (.08)	Best integration int. (.09)	Lowest ongoing cost (.08)	Total (across 14 criteria)	Relative Decimal Value
Program A	.03 (.01)	.33 (.03)	.02 (0)	.16	.18
Program B	.48 (.04)	.33 (.03)	.73 (.06)	.30	.33
Program C	.48 (.04)	.33 (.03)	.25 (.02)	.44	.49
			Grand Total	.90	

Information provided courtesy of Novacor Chemicals

Result: Program C was chosen. Even though 14 out of the 16 team members were not currently using this program, the prioritization process changed their minds, and prevented them from biasing the final decision.

Problem-Solving/ Process-Improvement Model

Improvement Storyboard

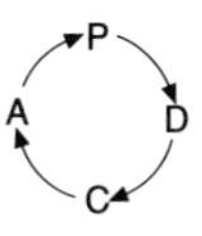

There are many standard models for making improvements. They all attempt to provide a repeatable set of steps that a team or individual can learn and follow. The Improvement Storyboard is only *one* of many models that include typical steps using typical tools. Follow this model or any other model that creates a common language for continuous improvement within your organization.

Plan

1. **Select the problem/process that will be addressed first (or next) and describe the improvement opportunity.**
2. **Describe the current process surrounding the improvement opportunity.**
3. **Describe all of the possible causes of the problem and agree on the root cause(s).**
4. **Develop an effective and workable solution and action plan, including targets for improvement.**

Do

5. **Implement the solution or process change.**

Check

6. **Review and evaluate the result of the change.**

Act

7. **Reflect and act on learnings.**

Plan

Depending on your formal process structure, Step 1 may be done by a steering committee, management team, or improvement team. If you are an improvement team leader or member, be prepared to start with Step 1 *or* Step 2.

1. **Select the problem/process that will be addressed first (or next) and describe the improvement opportunity.**
   - Look for changes in important business indicators
   - Assemble and support the right team
   - Review customer data
   - Narrow down project focus. Develop project purpose statement

Typical tools
Brainstorming, Affinity Diagram, Check Sheet, Control Chart, Histogram, Interrelationship Digraph, Pareto Chart, Prioritization Matrices, Process Capability, Radar Chart, Run Chart

Situation
Stop 'N Go Pizza* is a small but recently growing pizza delivery business with six shops. After a period of rapid growth, Stop 'N Go Pizza experienced a six-month decline in volume. Customers were leaving. Top management formed a mixed team of store managers, kitchen staff, and delivery personnel to find out why, and to generate an implementation plan to correct the situation. The team used both the Run Chart and Pareto Chart.

---

* The name Stop 'N Go Pizza, and the data associated with this case study are fictional. Any similiarity to an actual company by this name is purely coincidental.

Run Chart

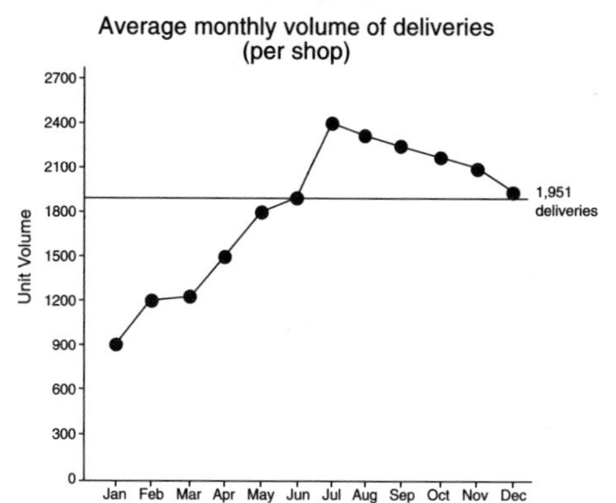

Pareto Chart

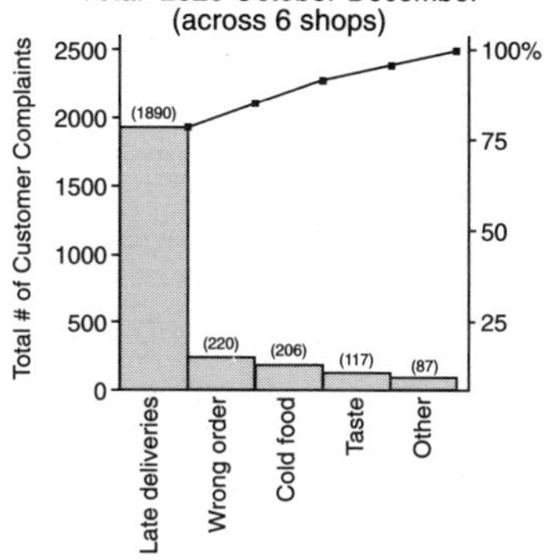

Illustration note: Delivery time was defined by the total time from when the order was placed to when the customer received it.

Pareto Chart

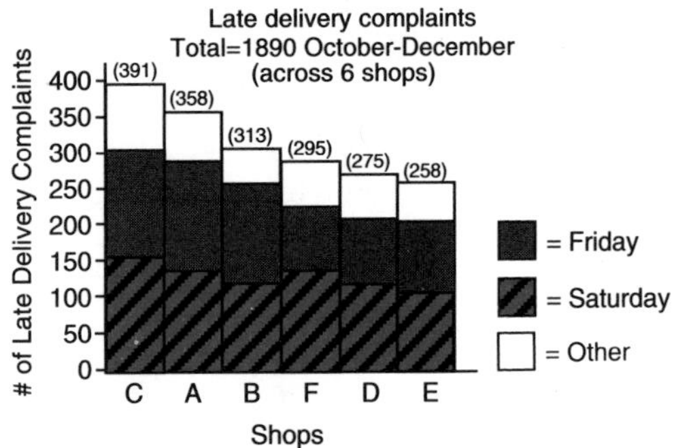

Decision
Late deliveries, (late from the time of order), were *by far* the most frequent customer complaint across all locations, especially on Fridays and Saturdays.

Team purpose statement
Reduce late deliveries on Fridays and Saturdays.

2. **Describe the current process surrounding the improvement opportunity.**
   - Select the relevant process or process segment to define the scope of the project
   - Describe the process under study

Typical tools
Brainstorming, Macro, Top-down, and Deployment Flowcharts, Tree Diagram

Situation
The team began to understand the overall process for producing and delivering their product and problems

that contributed to the project focus. The team used a Macro Flowchart.

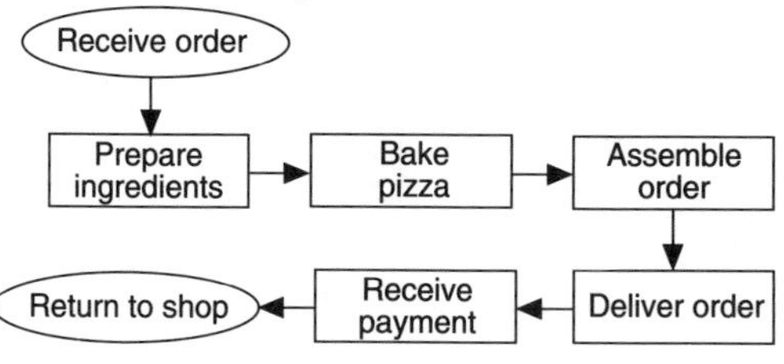

Decision

It became obvious that the "late deliveries" went far beyond the physical delivery process. Everything in the Macro Flowchart affected the "order-to-eating" time. This total process improvement became the team's focus.

3. **Describe all of the possible causes of the problem and agree on the root cause(s).**
   - Identify and gather helpful facts and opinions on the cause(s) of the problem
   - Confirm opinions on root cause(s) with data whenever possible

Typical tools

Affinity Diagram, Brainstorming, C & E / Fishbone Diagram, Check Sheet, Force Field Analysis, Interrelationship Digraph, Multivoting, Nominal Group Technique, Pareto Chart, Run Chart, Scatter Diagram

Situation

The team brainstormed all of the possible causes for "late deliveries" and then continued to ask "Why?" so that possible cause patterns could emerge. The team used a C & E/Fishbone Diagram and Run Charts.

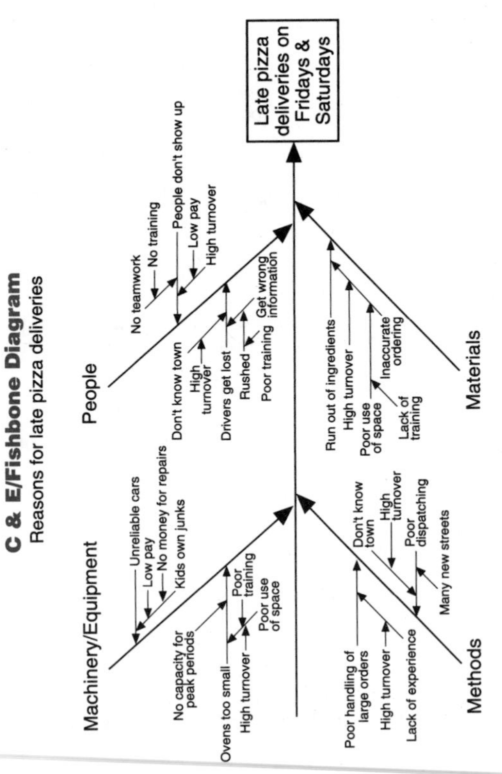

Run Charts

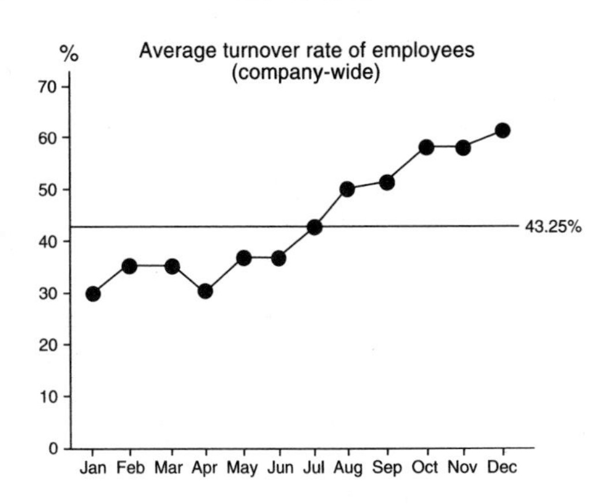

%
Average turnover rate of employees
(company-wide)
70
60
50
40
30
20
10
0
43.25%
Jan Feb Mar Apr May Jun Jul Aug Sep Oct Nov Dec

Average training hours of new employees
Average # of Hours
14
12
10
8
6
4
2
0
8 hours
Jan Feb Mar Apr May Jun Jul Aug Sep Oct Nov Dec

Decision
The C & E/Fishbone Diagram repeatedly pointed to "turnover" and "lack of training" as root causes. This applied to *ALL* areas of the operation, *NOT* just in the actual delivery portion of the process. Subsequent data showed that as the business rapidly grew, less time was put into training all employees. With this lack of adequate training, many employees felt a great deal of pressure at the busiest times. They also felt that they were unable to do their jobs well. This combination of work pressure and lack of self-confidence often caused employees to quit.

4. **Develop an effective and workable solution and action plan, including targets for improvement.**
   - Define and rank solutions
   - Plan the change process: What? Who? When?
   - Do contingency planning when dealing with new and risky plans
   - Set targets for improvement and establish monitoring methods

Typical tools
Activity Network Diagram, Brainstorming, Flowchart, Gantt Chart, Multivoting, Nominal Group Technique, PDPC, Prioritization Matrices, Matrix Diagram, Tree Diagram

Situation
The team used the combination of the Tree Diagram, Prioritization Matrices, Responsibility Matrix, and Gantt Chart to create a workable plan that attacked the heart of the problem.

Decision
The team focused on the most effective, efficient way to deliver the new training. They developed a four-month

Tree Diagram

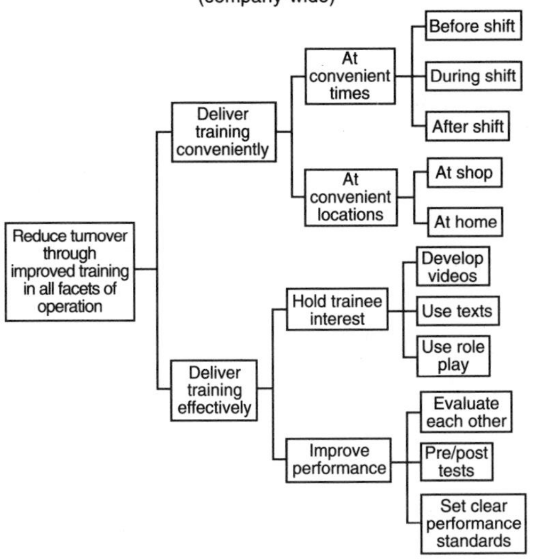

implementation plan that featured the creative use of videotapes, role plays, peer ratings, and so on. The team set the following targets based on past performance:

- Reduce turnover rate from 62 to 30 percent
- Reduce average time of order-to-delivery from 40 to 25 minutes
- Reduce customer complaints of late deliveries by 50 percent, without increasing other complaint categories
- Increase average monthly volume to 2400 units per shop from the current 1891 units

Prioritization Matrix

Selecting the best training program components

Tasks & Options \ Criteria & Weighting	Effectiveness (.60)	Feasibility (.19)	Time (.19)	Cost (.01)	Total
Train before shift	⦿	⦿	○	○	7.70
Train during shift	○	△	⦿	⦿	3.78
Train after shift	△	○	○	○	1.77
Train at the shop	⦿	⦿	○	○	7.70
Train at home	○	⦿	⦿	⦿	5.29
Develop videos	⦿	⦿	⦿	○	8.83
Use texts	○	⦿	○	⦿	4.16
Use role play	⦿	⦿	⦿	⦿	8.89
Evaluate each other	⦿	⦿	○	○	7.70
Pre/post test	⦿	⦿	⦿	○	8.83
Set clear performance standards	⦿	⦿	⦿	⦿	8.89

⦿ = 9 Excellent ○ = 3 Fair △ = 1 Poor

The total = the sum of [rating values x criteria weighting]

For example, to find the total of the "Train before shift" row, do the following:

[⦿ (9) x .60] + [⦿ (9) x .19] + [○ (3) x .19] + [○ (3) x .01] = 7.70

Note: **Weighting values** of each criterion come from a criteria matrix not shown.

Task options come from the most detailed level of the Tree Diagram on the previous page.

Matrix & Gantt Chart Combined

New training program timeline

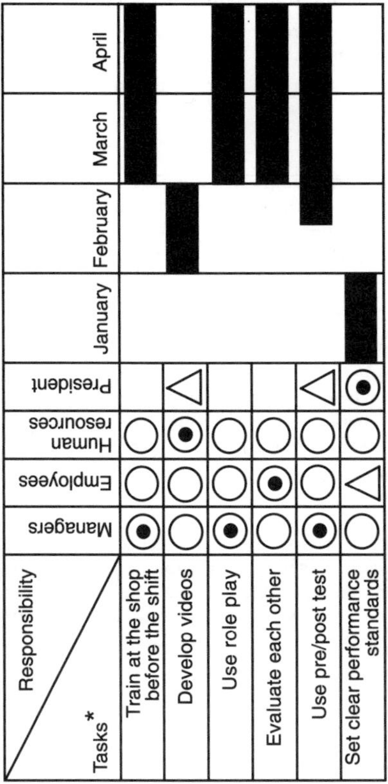

Tasks* / Responsibility	Managers	Employees	Human resources	President	January	February	March	April
Train at the shop before the shift	◉	○	○				■	■
Develop videos	○	○	◉	△		■		
Use role play	◉	○	○				■	■
Evaluate each other	○	◉	○				■	■
Use pre/post test	◉	○	○	△		■	■	■
Set clear performance standards	○	△	○	◉	■			

◉ = Primary responsibility

○ = Secondary/team member

△ = Need information to/from

*These were the highest rated tasks from the Prioritization Matrix on the previous page.

Do

5. **Implement the solution or process change.**
   - It is often recommended to try the solution on a small scale first
   - Follow the plan and monitor the milestones and measures

Typical tools

Activity Network Diagram, Flowchart, Gantt Chart, Matrix Diagram, and other project management methods, as well as gathering ongoing data with Run Charts, Check Sheets, Histograms, Process Capability, and Control Charts

Situation

The team used the Responsibility Matrix and Gantt Chart to guide the training rollout. The original plan was to start in only two shops. There were some initial problems with employees not being paid for the training time as well as some managers not getting coverage during training.

Decision

The solution was to pay the employees for one-half the training time and to set up overlapping shifts for better coverage for managers. Once these adjustments were made, the training went so well and had so many unintended benefits that it was expanded to all six shops.

Check

6. Review and evaluate the result of the change.

- Confirm or establish the means of monitoring the solution. Are the measures valid?
- Is the solution having the intended effect? Any unintended consequences?

Typical tools
Check Sheet, Control Chart, Flowchart, Pareto Chart, Run Chart

Situation
The training plan was developed in January and February and rolled out in March and April. The team went back to the original Pareto Chart data to compare it to the current customer complaint data. In addition, they collected current turnover and delivery time measures to test the original connection with improved training.

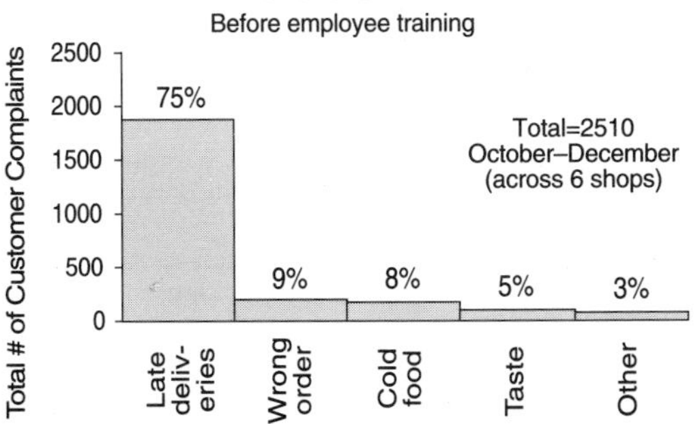

The Pareto Chart on the next page shows the types of customer complaints after the training plan was rolled out in March and April.

Pareto Chart

After employee training

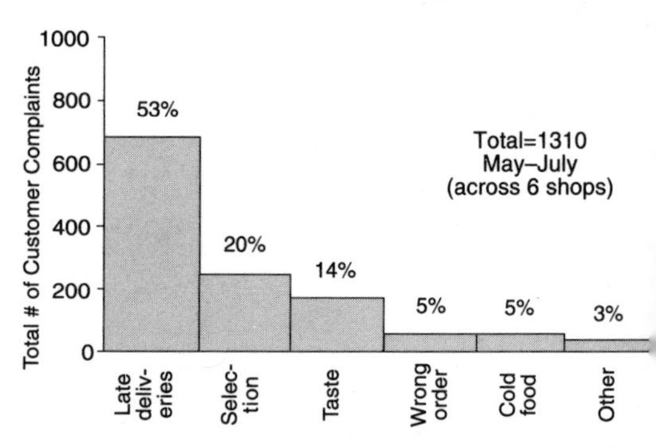

Decision

The team reviewed the original targets, which were set in Step 4.

Turnover: While not at the 30 percent average, it had decreased for six consecutive months from a high of 62 percent to 44 percent last month.

Delivery Time: Steadily declined from a high of 40 minutes to its most recent level of 28 minutes.

Customer Complaints: Overall, complaints were reduced by 52 percent and within "late deliveries" by 63 percent

Sales Volume: The average volume last month was at 2250 units, up for the third straight month.

Unexpected result

For the first time, customers complained about the lack of a good selection on the menu. While Stop 'N Go Pizza was working toward increasing its speed of pizza deliveries and standardizing its processes, the variety of the menu was perhaps too simplified.

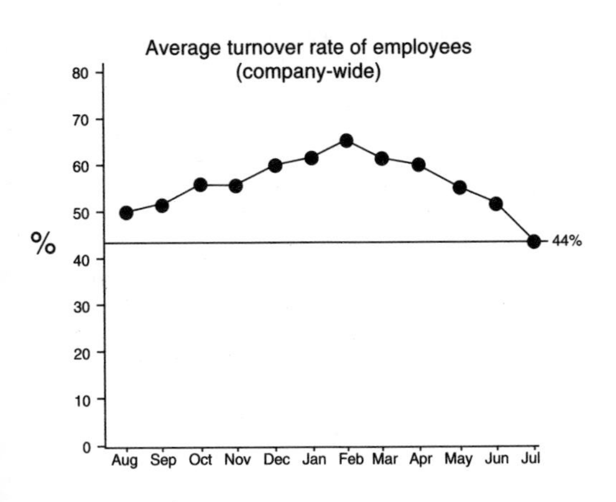

Average turnover rate of employees
(company-wide)
%
80
70
60
50
40
30
20
10
0
44%
Aug Sep Oct Nov Dec Jan Feb Mar Apr May Jun Jul

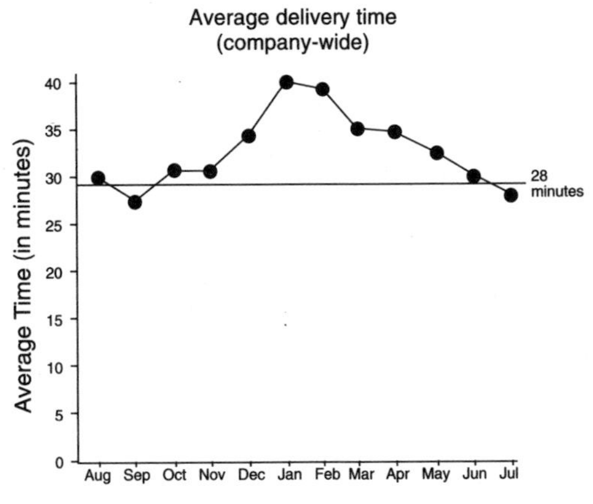

Average delivery time
(company-wide)
Average Time (in minutes)
40
35
30
25
20
15
10
5
0
28
minutes
Aug Sep Oct Nov Dec Jan Feb Mar Apr May Jun Jul

Act

7. **Reflect and act on learnings.**
   - Assess the results and problem-solving process and recommend changes
   - Continue the improvement process where needed; standardization where possible
   - Celebrate success

Typical tools
Affinity Diagram, Brainstorming, Improvement Storyboard, Radar Chart

Situation
Six months after the new training was started, the team met to evaluate its results and process. Team members used a Radar Chart to illustrate their assessment of the team.

Decision
The Radar Chart showed strong agreement among team members on the performance of "Results," "Use of tools," and "Impact on customers." The performance and consensus among team members were both lower in "Standardization" and "Teamwork." When the team presented its storyboard to top management, the major result was a complete overhaul of basic training content and delivery, as well as the new performance measures that would continue to be monitored. The final celebration was... what else... an all-expense-paid Stop 'N Go Pizza party!

Future Possibilities
More efficient mapping, routing, and dispatching of pizza deliveries, as well as more staff cross-training.

Radar Chart

Team evaluation of itself after new training

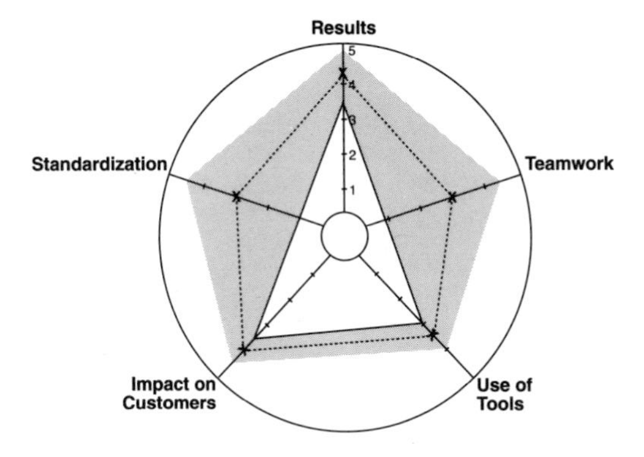

Note: The "x" mark indicates the team's average performance rating while the shaded area indicates the range of ratings within the team.

Process Capability

Measuring conformance to customer requirements

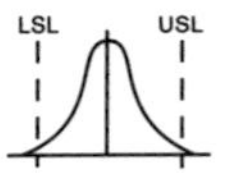

Why use it?

To determine whether a process, given its natural variation, is capable of meeting established customer requirements or specifications.

What does it do?

- Helps a team answer the question, "Is the process capable?"
- Helps to determine if there has been a change in the process
- Helps to determine percent of product or service not meeting customer requirements

How do I do it?

1. **Determine the process grand average, $\bar{\bar{X}}$, and the average range, $\bar{R}$**
   - Use a stable Control Chart, which means the process is stable and normally distributed.

2. **Determine the Upper Specification Limit (USL) and the Lower Specification Limit (LSL)**
   - The USL and LSL are based on customer requirements. Recognize that these specification limits are based *solely* on customer requirements and do not reflect the capacity of the process.

3. **Calculate the process standard deviation**
   - Process cabability is based on individual points from a process under study. Information from a Control Chart can be used to estimate the process'

average and variation (standard deviation, s).

- σ is a measure of the process (population) standard deviation and can be estimated from information on the Control Chart by

$$\hat{\sigma} = \frac{\bar{R}}{d_2} \text{ or } \hat{\sigma} = \frac{\bar{s}}{c_4}$$

where $\bar{R}$ and $\bar{s}$ are the averages of the subgroup ranges and standard deviation, and d_2 and c_4 are the associated constant values based on the subgroup sample sizes. See the Table of Constants in Control Charts.

- The process average is estimated simply by $\bar{\bar{X}}$, $\bar{\tilde{X}}$, $\bar{X}$.

4. **Calculate the process capability**

- To measure the degree to which a process is or is not capable of meeting customer requirements, capability indices have been developed to compare the distribution of your process in relation to the specification limits.
- A stable process can be represented by a measure of its variation—six standard deviations. Comparing six standard deviations of the process variation to the customer specifications provides a measure of capability. Some measures of capability include C_p and its inverse C_r, C_{pl}, C_{pu}, and C_{pk}.

C_p (simple process capability)

$$C_p = \frac{USL - LSL}{6\hat{\sigma}}$$

Tip While C_p relates the spread of the process relative to the specification width, it *DOES NOT* look at how well the process average is centered to the target value.

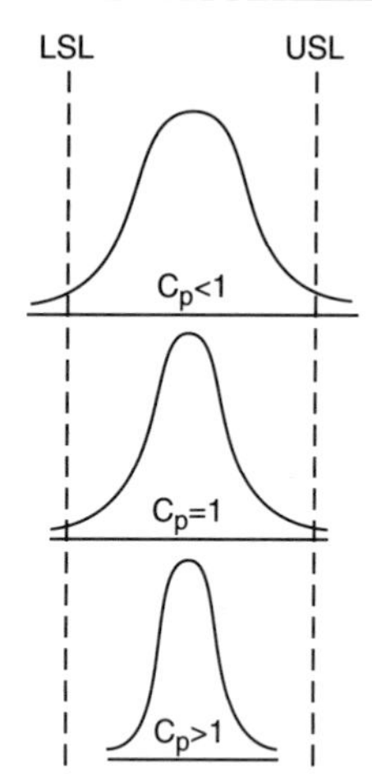

The process variation exceeds specification. Defectives are being made.

The process is just meeting specification. A minimum of .3% defectives will be made, more if the process is not centered.

The process variation is less than specification, however, defectives might be made if the process is not centered on the target value.

C_{pl}, C_{pu}, and C_{pk} (process capability indices)

The indices C_{pl} and C_{pu} (for single-sided specificiation limits) and C_{pk} (for two-sided specification limits) measure not only the process variation with respect to the allowable specification, they also take into account the location of the process average. C_{pk} is considered a measure of the process capability and is taken as the smaller of either C_{pl} or C_{pu}

$$C_{pl} = \frac{\overline{\overline{X}} - LSL}{3\hat{\sigma}} \quad C_{pu} = \frac{USL - \overline{\overline{X}}}{3\hat{\sigma}} \quad C_{pk} = \min\{C_{pl}, C_{pu}\}$$

Tip If the process is near normal and in statistical control, C_{pk} can be used to estimate the expected percent of defective material. Estimating the percentages of defective material is beyond the scope of this book and can be found in statistical books.

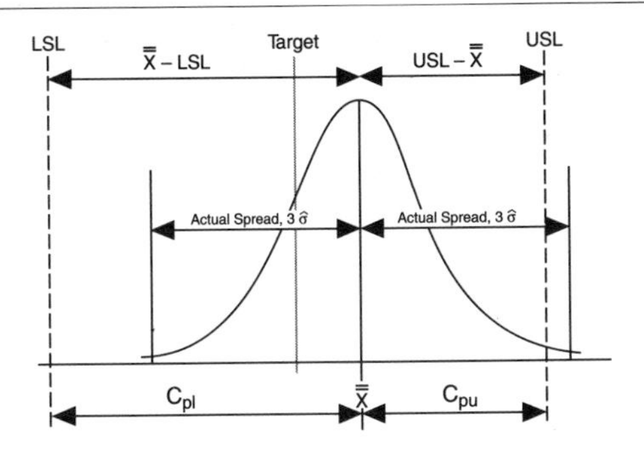

- If the process is not capable, form a team to identify and correct the common causes of the variation in the process.
  - Process capability, based on individual data of the process population, is used to determine if a process is capable of meeting customer requirements or specifications. It represents a "snapshot" of the process for some specific period of time.
  - Control Charts use small sample sizes over time and look at the averages. The control limits are natural limits of variation of the *averages within the sample*. These limits are not to be confused with specification limits, which are for *individual data points in the population.*

Variations

The construction steps described in this section are based on process capability of a Variable Data Control Chart. The process capability of an Attribute Data Control Chart is represented by the process averages $\bar{p}$, $n\bar{p}$, $\bar{c}$, and $\bar{u}$.

Process Capability

Die Cutting Process

A Control Chart was maintained, producing the following statistics:

$\bar{\bar{X}} = 212.5 \quad \bar{R} = 1.2 \quad n = 5$

Spec. = 210 ± 3 USL = 213 LSL = 207

$$\hat{\sigma} = \frac{\bar{R}}{d_2} = \frac{1.2}{2.326} = .516$$

$$C_p = \frac{USL - LSL}{6\hat{\sigma}} = \frac{213 - 207}{6(.516)} = \frac{6}{3.096} = 1.938$$

$$C_{pl} = \frac{\bar{\bar{X}} - LSL}{3\hat{\sigma}} = \frac{212.5 - 207}{3(.516)} = \frac{5.5}{1.548} = 3.553$$

$$C_{pu} = \frac{USL - \bar{\bar{X}}}{3\hat{\sigma}} = \frac{213 - 212.5}{3(.516)} = \frac{0.5}{1.548} = 0.323$$

$$C_{pk} = \min\{C_{pl}, C_{pu}\} = 0.323$$

Since $C_{pk}<1$, defective material is being made.

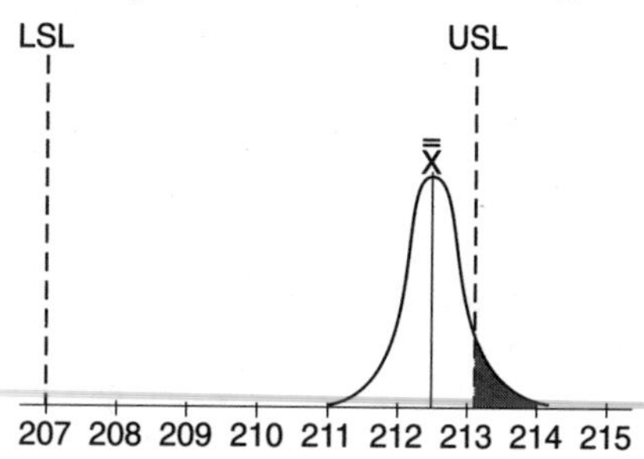

Radar Chart

Rating organization performance

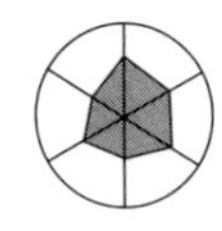

Why use it?

To visually show in one graphic the size of the gaps among a number of *both current* organization performance areas *and ideal* performance areas.

What does it do?

- Makes concentrations of strengths and weaknesses visible
- Clearly displays the important categories of performance
- If done well, clearly defines full performance in each category
- Captures the different perceptions of all the team members about organization performance

How do I do it?

1. **Assemble the right team/raters**

   Tip It is critical to get varied perspectives to avoid organization "blind spots."

2. **Select and define the rating categories**
   - The chart can handle a wide number of categories, with 5–10 categories as an average.
   - Brainstorm or bring headers from an Affinity Diagram to create the categories.
   - Define both non-performance and full performance within each category so ratings are done consistently.

3. Construct the chart

- Draw a large wheel on a flipchart with as many spokes as there are rating categories.
- Write down each rating category at the end of each spoke around the perimeter of the wheel.
- Mark each spoke on a zero to "n" scale with "0" at the center equal to "no performance" and the highest number on the scale at the outer ring equal to "full performance." Performance can be measured either objectively or subjectively.

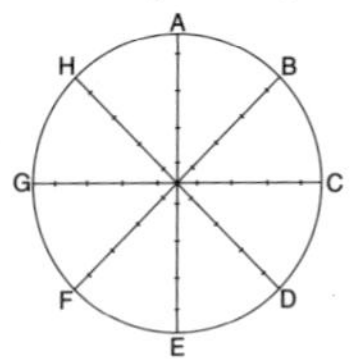

4. Rate all performance categories

a) **Individual:** Each person rates in silence, using multicolored markers or adhesive labels directly on the flipchart.

b) **Team:** Through consensus or an average of individual scores, get a team rating. Take into account both the clustering and the spread of the individual ratings.

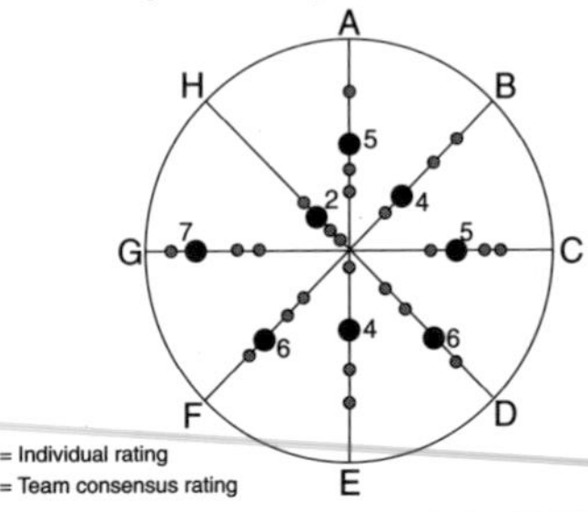

Tip Make the team rating highly visible on the chart. Be sure to differentiate the team ratings from individual ratings on the chart by color or type of mark.

5. **Connect the team ratings for each category and highlight as needed**

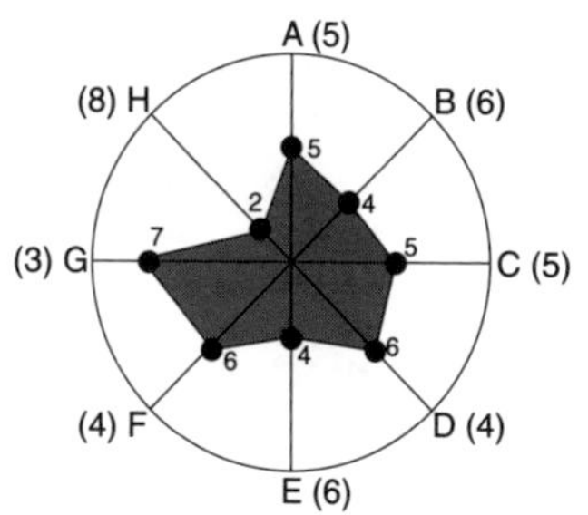

Gap scores are in parentheses.

Tip A gap score can be added to each category by subtracting the team rating score from the highest number on the rating scale, e.g., on a scale of "10," a team rating of "4" produces a gap score of "6" in categories B and E.

6. **Interpret and use the results**
   - The overall ratings identify gaps within each category but not the relative importance of the categories themselves. Work on the biggest gap in the most critical ***category.***
   - Post the resulting Radar Chart in a prominent place, review progress regularly, and update the chart accordingly. It is a great visual "report card."

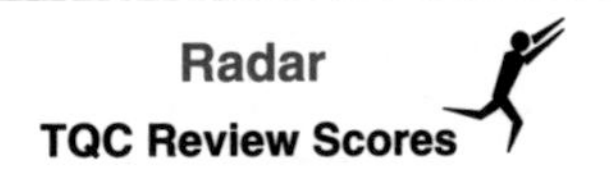

Radar

TQC Review Scores

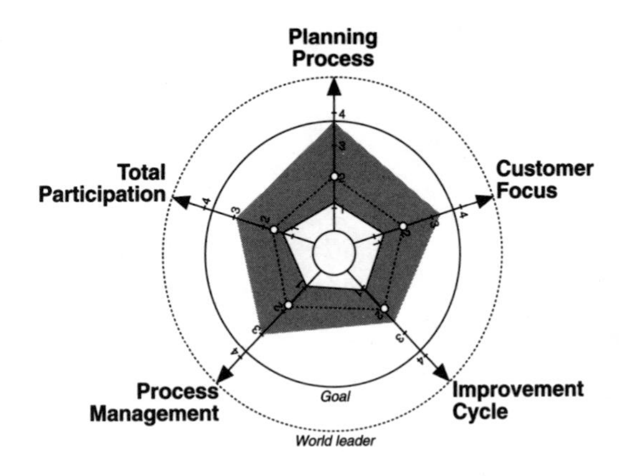

Range of ratings within the team

o···o Average

Company's goal: to have 80% of all entities (34) achieve an overall score of >3.5.

To compute overall score:

$$\frac{\text{Sum of average scores from each category}}{\text{\# of categories}} = \frac{12.52}{5} = 2.5 \text{ (maximum is 5)}$$

Information provided courtesy of Hewlett Packard

Run Chart

Tracking trends

Why use it?

To allow a team to study observed data (a performance measure of a process) for trends or patterns over a specified period of time.

What does it do?

- Monitors the performance of one or more processes over time to detect trends, shifts, or cycles
- Allows a team to compare a performance measure before and after implementation of a solution to measure its impact
- Focuses attention on truly vital changes in the process
- Tracks useful information for predicting trends

How do I do it?

1. **Decide on the process performance measure**

2. **Gather data**
   - Generally, collect 20-25 data points to detect meaningful patterns.

3. **Create a graph with a vertical line (y axis) and a horizontal line (x axis)**
   - On the vertical line (y axis), draw the scale related to the variable you are measuring.
     – Arrange the y axis to cover the full range of the measurements and then some, e.g., $1\frac{1}{2}$ times the range of data.

- On the horizontal line (x axis), draw the time or sequence scale.

4. **Plot the data**
   - Look at the data collected. If there are no obvious trends, calculate the average or arithmetic mean. The average is the sum of the measured values divided by the number of data points. The median value can also be used but the mean is the most frequently used measure of the "centering" of the sample. (See Data Points for more information on averages.) Draw a horizontal line at the average value.

Tip Do not redraw this average line every time new data is added. Only when there has been a significant change in the process or prevailing conditions should the average be recalculated and redrawn, and then only using the data points after the verified change.

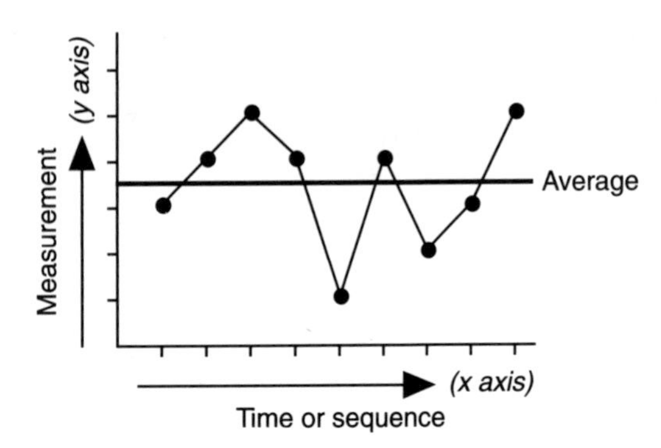

5. **Interpret the Chart**

- Note the position of the average line. Is it where it should be relative to a customer need or specification? Is it where you want it relative to your business objective?

Tip A danger in using a Run Chart is the tendency to see every variation in data as being important. The Run Chart should be used to focus on truly vital changes in the process. Simple tests can be used to look for meaningful trends and patterns. These tests are found in Control Charts in the "Determining if Your Process is Out of Control" section. Remember that for more sophisticated uses, a Control Chart is invaluable since it is simply a Run Chart with statistically-based limits.

Run

Average Number of Days for Determining Eligibility for Services

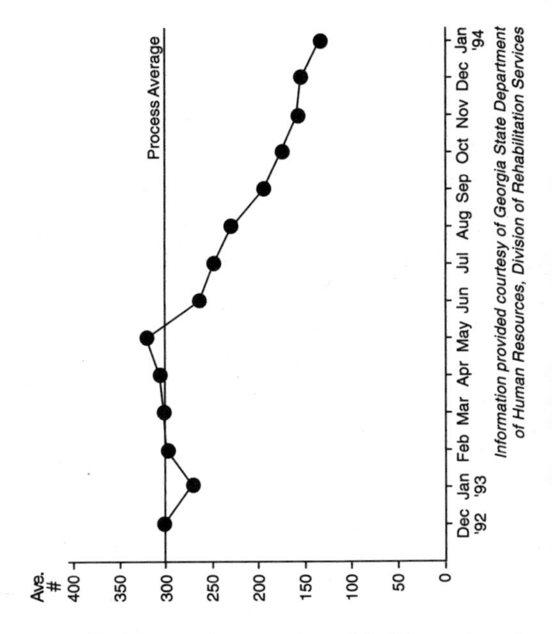

Note: Eligibility requirements changed in May, making it much simpler for the department staff to make determinations. The trend is statistically significant because there are six or more consecutive points declining.

Scatter Diagram

Measuring relationships between variables

Why use it?

To study and identify the possible relationship between the changes observed in two different sets of variables.

What does it do?

- Supplies the data to confirm a hypothesis that two variables are related
- Provides both a visual and statistical means to test the strength of a potential relationship
- Provides a good follow-up to a Cause & Effect Diagram to find out if there is more than just a consensus connection between causes and the effect

How do I do it?

1. Collect 50–100 paired samples of data that you think may be related and construct a data sheet

Course	Average Session Rating (on a 1–5 scale)	Average Experience of Training Team (days)
1	4.2	220
2	3.7	270
3	4.3	270
•	•	•
•	•	•
•	•	•
40	3.9	625

Theory: There is a possible relationship between the number of days of experience the training team has received and the ratings of course sessions.

2. **Draw the horizontal (x axis) and vertical (y axis) lines of the diagram**
   - The measurement scales generally increase as you move up the vertical axis and to the right on the horizontal axis.

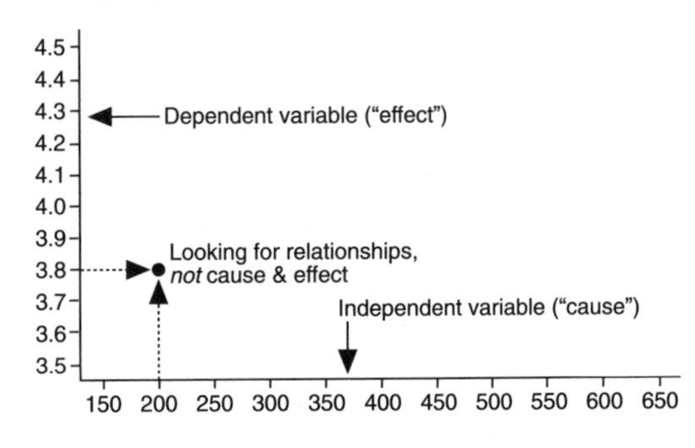

3. **Plot the data on the diagram**
   - If values are repeated, circle that point as many times as appropriate.

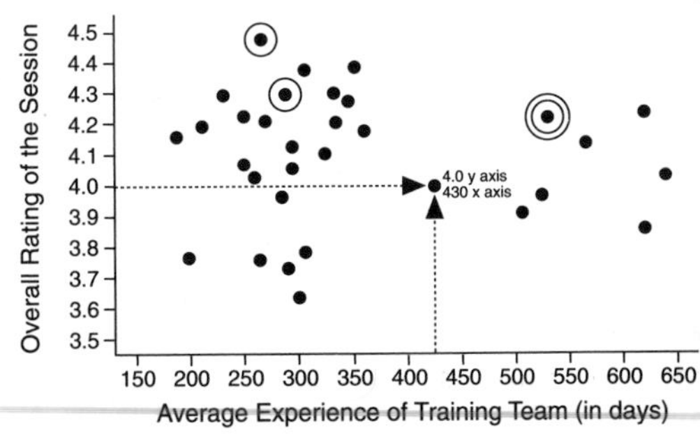

Information provided courtesy of Hamilton Standard

4. **Interpret the data**

- There are many levels of analysis that can be applied to Scatter Diagram data. Any basic statistical process control text, like Kaoru Ishikawa's *Guide to Quality Control*, describes additional correlation tests. It is important to note that all of the examples in this chapter are based on straight-line correlations. There are a number of non-linear patterns that can be routinely encountered, e.g., $y = e^x$, $y = x^2$). These types of analyses are beyond the scope of this book.
- The following five illustrations show the various patterns and meanings that Scatter Diagrams can have. The example used is the training session assessment previously shown. The patterns have been altered for illustrative purposes. Pattern #3 is the actual sample.

Tip The Scatter Diagram *does not predict* cause and effect relationships. It only shows the strength of the relationship between two variables. The stronger the relationship, the greater the likelihood that change in one variable will affect change in another variable.

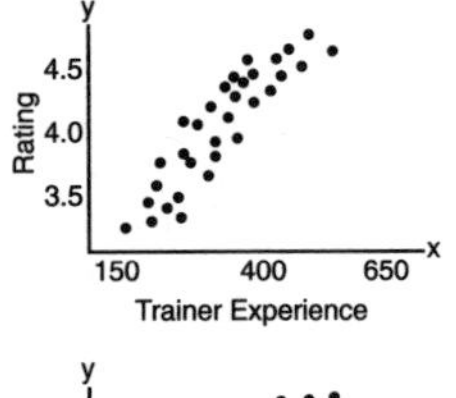

1. Positive Correlation. An increase in y may depend on an increase in x. Session ratings are likely to increase as trainer experience increases.

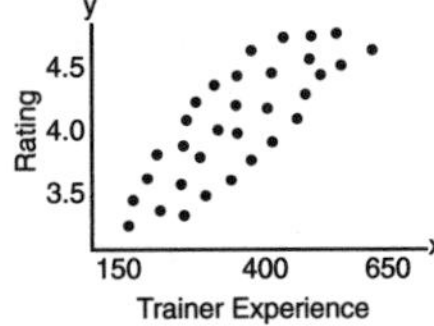

2. Possible Positive Correlation. If x is increased, y may increase somewhat. Other variables may be involved in the level of rating in addition to trainer experience.

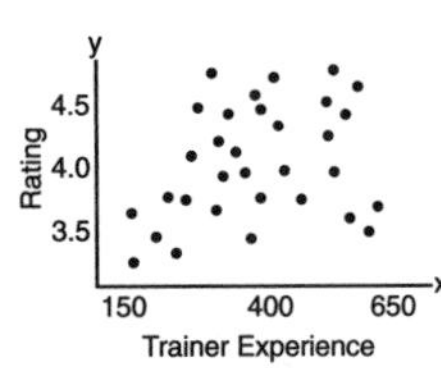

3. No Correlation. There is no demonstrated connection between trainer experience and session ratings.

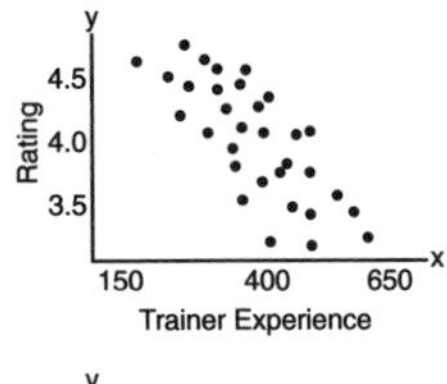

4. Possible Negative Correlation. As x is increased, y may decrease somewhat. Other variables, besides trainer experience, may also be affecting ratings.

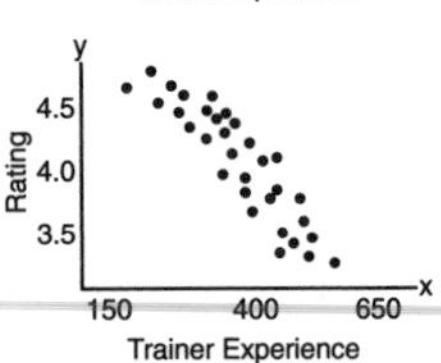

5. Negative Correlation. A decrease in y may depend on an increase in x. Session ratings are likely to fall as trainer experience increases.

Scatter

Capacitance vs. Line Width

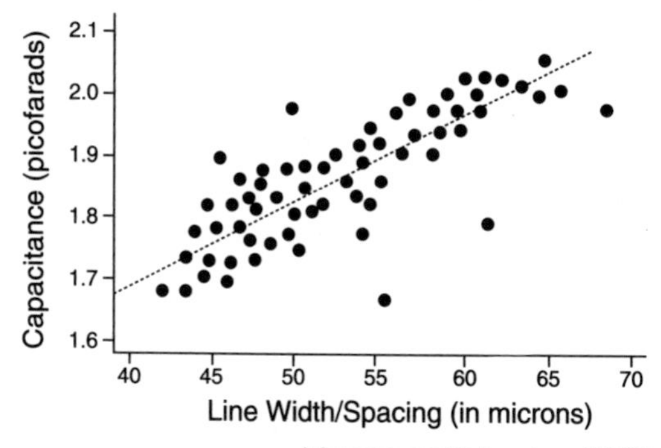

Information provided courtesy of AT&T

Note: This Scatter Diagram shows that there is a strong positive relationship between these two variables in producing microelectronic circuits. Since capacitance measures a critical performance of a circuit, anything that affects it positively or negatively is also critical. The diagram shows that line width/spacing is something to watch closely, perhaps using a Control Chart or another type of statistical process control (SPC) tool.

Team Guidelines

From "me" to "we"

Starting Teams

The most critical task for any new team is to establish its purpose, process, and measures of team progress. Once the team has developed the following guidelines and charters specific to its purpose, they should be recorded on a flipchart and posted at each team meeting for reference.

- **Develop a Team Behavior Charter**
  - *Groundrules.* Develop consensus groundrules of acceptable and unacceptable individual and team behavior.
  - *Decision making.* Determine whether decisions will be made by consensus, majority rule, or anarchy! Discuss whether there are, or should be, exceptions to when the team should not follow its usual process.
  - *Communication.* Recognize the value of listening and constructive feedback, and make the effort, every day, to communicate constructively!
  - *Roles and participation.* Discuss how the team will choose a leader, and generally how the team process will be led. The individuals and team must take responsibility to encourage equal participation.
  - *Values.* Acknowledge and accept the unique insight of each member of the team.

- **Develop a Purpose Charter**
  - Establish the answer to why the team exists.

– Bring together the individuals who would work well together as a team. Determine whether each person has the knowledge, skills, and influence required to participate effectively on the team.
– The team should discuss who its customers are. If the team has multiple customers, decide which customers have the highest priority, or at least how their needs will be balanced.

- **Develop Measures of Team Progress**
  – Discuss and agree on the desired signals, which the team can assess both objectively and subjectively, that will indicate the team is making progress.
  – Discuss and agree on the types of measures and outcomes that will indicate the team has reached success or failure.
  – Estimate the date when the project should be completed.

Maintaining Momentum

Many teams enjoy terrific starts and then soon fizzle. The real challenge is to keep a team focused on its purpose and not the histories of its members and their relationships to one another.

- **Agree on the Improvement Model to Use**
  – *Standard steps.* Use your organization's standard step-by-step improvement process or choose from the many published options. (See the Improvement Storyboard in the Problem Solving/Process Improvement Model section for one such standard process.)

– *Data.* Gather relevant data to analyze the current situation. Define what you know, and what you need to know, but know when to stop. Learn, as a team, to say when your work is good enough to proceed to the next step in the process.
– *Develop a plan.* Use your organization's standard improvement model to provide the overall structure of a project plan. Estimate times for each step and for the overall project. Monitor and revise the plans as needed.

- **Use Proven Methods Based on Both Data and Knowledge**
  – *Data-based methods.* Use tools in this booklet, e.g., Run Chart, Pareto Chart, that reveal patterns within data. These tools often take the emotion out of discussions and keep the process moving.
  – *Knowledge-based methods.* Many of the methods in this booklet, e.g., Affinity Diagram, Interrelationship Digraph, help to generate and analyze ideas to reveal the important information within. They help create consensus, which is the ideal energy source for a team.

- **Manage Team Dynamics**
  – *Use facilitators.* A facilitator is someone who monitors and helps team members to keep their interactions positive and productive. This is the stage when a facilitator can help the team stay focused on its purpose while improving its working relationship.

- *Manage conflict*. As teams grow, so do conflicts. This is a natural process as communication becomes more open. The entire team can learn techniques for conflict resolution and use the facilitator as a resource.
- *Recognize agreement*. Managing agreement is often as much of an effort as managing disagreement. Test for agreement often and write down the points of agreement as they occur.
- *Encourage fair participation*. Each team member must eventually take responsibility for participating consistently in all discussions. Likewise, the entire team should be constantly working to "pull back" the dominant members and draw out the quieter members.

Ending Teams/Projects

Most teams and all projects must eventually end. Both often end in unsatisfactory ways or don't "officially end" at all. Before ending, the team should review the following checklist:

- ❐ We checked our results against our original goals and customer needs.
- ❐ We identified any remaining tasks to be done.
- ❐ We established responsibility for monitoring the change over time.
- ❐ We documented and trained people, when necessary, in the new process.

❐ We communicated the changes to everyone affected by them.

❐ We reviewed our own team's accomplishments for areas of improvement.

❐ We celebrated the efforts of the team with a lunch, newsletter article, special presentation to the company, or other expression of celebration.

❐ We feel proud of our contribution and accomplishments, our new capabilities, and our newly defined relationships with coworkers.

Conducting Effective Meetings

Preparation:

- Decide on the purpose of the meeting
- Develop a meeting plan (who, what, where, when, how many)
- Identify the meeting leader
- Prepare and distribute the agenda
- Set up the meeting area

Beginning:

- Start on time
- Introduce the meeting leader
- Allow team members to introduce themselves
- Ask for a volunteer timekeeper
- Ask for a volunteer recorder
- Review, change, order the agenda
- Establish time limits
- Review prior meeting action items

Meeting Etiquette:

- Raise your hand and be recognized before speaking
- Be brief and to the point
- Make your point calmly
- Keep an open mind
- Listen without bias
- Understand what is said
- Avoid side conversations
- Respect other opinions
- Avoid personal agendas
- Come prepared to do what's good for the company
- Have fun

Ending:

- Develop action items (who, what, when, how)
- Summarize the meeting with the group
- Establish the date and time for a follow-up meeting
- Evaluate the meeting
- End on time
- Clean the meeting area

Next Steps:

- Prepare and distribute the meeting activity report
- Follow up on action items
- Go to "Preparation"

Tree Diagram

Mapping the tasks for implementation

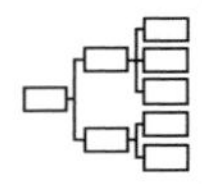

Why use it?

To break any broad goal, graphically, into increasing levels of detailed actions that must or could be done to achieve the stated goals.

What does it do?

- Encourages team members to expand their thinking when creating solutions. Simultaneously, this tool keeps everyone linked to the overall goals and subgoals of a task
- Allows all participants, (and reviewers outside the team), to check all of the logical links and completeness at every level of plan detail
- Moves the planning team from theory to the real world
- Reveals the *real* level of complexity involved in the achievement of any goal, making potentially overwhelming projects manageable, as well as uncovering unknown complexity

How do I do it?

1. Choose the Tree Diagram goal statement

Goal: Increase workplace suggestions

- Typical sources:
  - The root cause/driver identified in an Interrelationship Digraph (ID)
  - An Affinity Diagram with the headers as major subgoals
  - Any assignment given to an individual or team
- When used in conjunction with other management and planning tools, the most typical source is the root cause/driver identified in the ID.

Tip Regardless of the source, work hard to create—through consensus—a clear, action-oriented statement.

2. **Assemble the right team**
   - The team should consist of action planners with detailed knowledge of the goal topic. The team should take the Tree only to the level of detail that the team's knowledge will allow. Be prepared to hand further details to others.
   - 4-6 people is the ideal group size but the Tree Diagram is appropriate for larger groups as long as the ideas are visible and the session is well facilitated.

3. **Generate the major Tree headings, which are the major subgoals to pursue**
   - The simplest method for creating the highest, or first level of detail, is to brainstorm the major task areas. These are the major "means" by which the goal statement will be achieved.

- To encourage creativity, it is often helpful to do an "Action Affinity" on the goal statement. Brainstorm action statements and sort into groupings, but spend less time than usual refining the header cards. Use the header cards as the Tree's first-level subgoals.

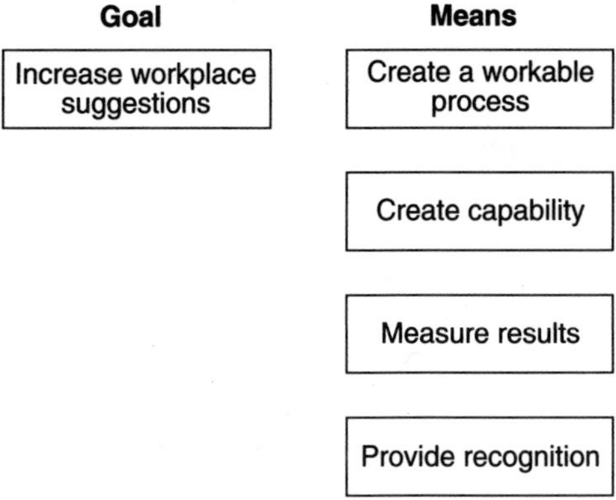

Goal	Means
Increase workplace suggestions	Create a workable process
	Create capability
	Measure results
	Provide recognition

Tip Use Post-it™ notes to create the levels of detail. Draw lines only when the Tree is finished. This allows it to stay flexible until the process is finished. The Tree can be oriented from left to right, right to left, or top down.

Tip Keep the first level of detail broad, and avoid jumping to the lowest level of task. Remember: "If you start with what you already know, you'll end up where you've already been."

4. **Break each major heading into greater detail**
   - Working from the goal statement and first-level detail, placed either to the extreme left, right or top of the work surface, ask of each first-level item:

     "What needs to be addressed to achieve the goal statement?"

     Repeat this question for each successive level of detail.
   - Stop the breakdown of each level when there are assignable tasks or the team reaches the limit to its own expertise. Most Trees are broken out to the third level of detail (not counting the overall goal statement as a level). However, some subgoals are just simpler than others and don't require as much breakdown.

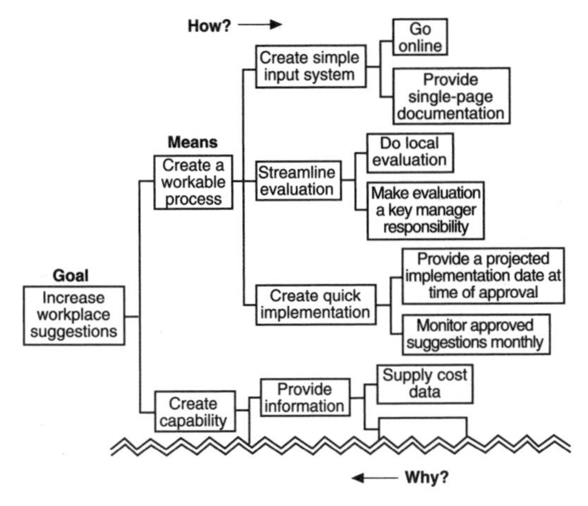

5. **Review the completed Tree Diagram for logical flow and completeness**
   - At each level of detail, ask "Is there something obvious that we have forgotten?"
   - As the Tree breaks down into greater detail (from general to specific) ask, "If I want to accomplish these results, do I really need to do these tasks?"
   - As the Tree builds into broader goals (from the specific to the general) ask, "Will these actions actually lead to these results?"
   - Draw the lines connecting the tasks.

   Tip The Tree Diagram is a great communication tool. It can be used to get input from those outside the team. The team's final task is to consider proposed changes, additions or deletions, and to modify the Tree as appropriate.

Variations

The Process Decision Program Chart (PDPC) is a valuable tool for improving implementation through contingency planning. The PDPC, based on the Tree Diagram, involves a few simple steps.

1. **Assemble a team closest to the implementation**

2. **Determine proposed implementation steps**
   - List 4–10 broad steps and place them in sequence in the first Tree level.

3. **Branch likely problems off each step**
   - Ask "What could go wrong?"

4. **Branch possible and reasonable responses off each likely problem**

5. Choose the most effective countermeasures and build them into a revised plan

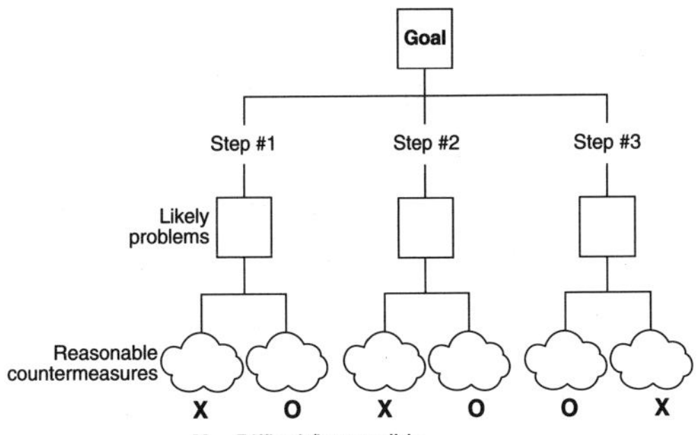

PDPC (Tree Variation)

Awarding Unrestricted Financial Aid

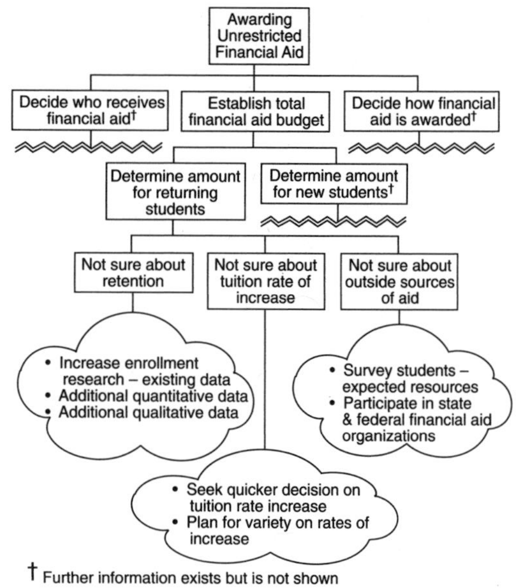

† Further information exists but is not shown

Information provided courtesy of St. John Fisher College

Note: The PDPC surfaced a lack of accurate information as a major problem. By anticipating this and filling the most critical information gaps, the budget can be more accurate.

Tree

Improve Business Planning Interaction

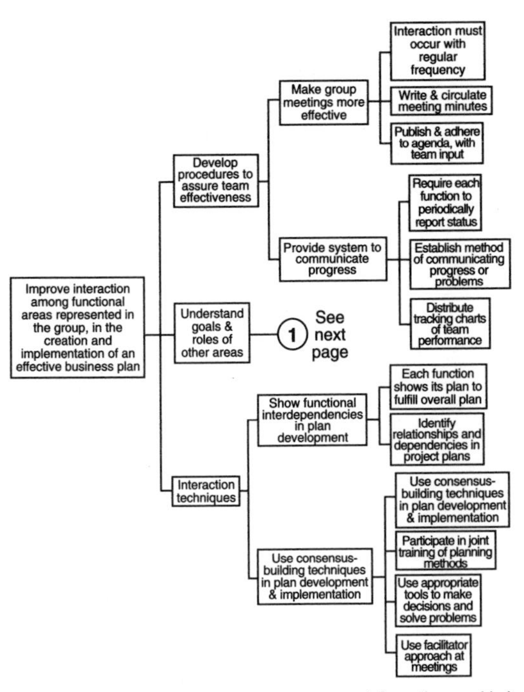

Information provided courtesy of Goodyear

Tree

Improve Business Planning Interaction (continued)

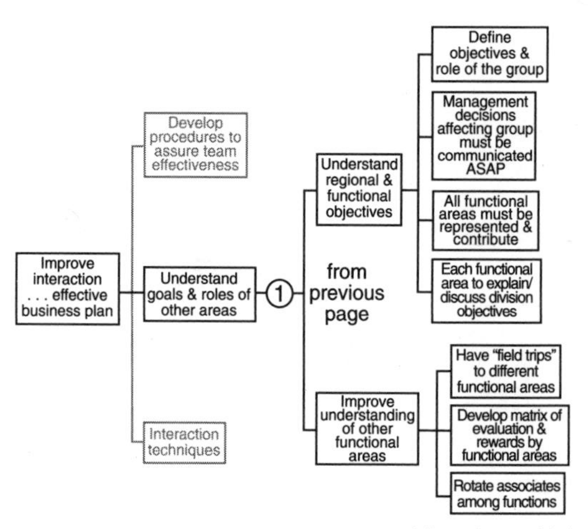

Information provided courtesy of Goodyear

The Memory Jogger™ 9000

The *Memory Jogger™ 9000* is the best source for you and everyone in your organization to learn how to comply with the ISO 9000 Standard and QS-9000 Requirements. It will show you how to: participate in identifying the work that you do that affects the quality of your organization's products and services; improve your work processes; write procedures that describe your work; and much more.

Code: 1060E Price: ***$7.95***

The Team Memory Jogger™

Easy to read and written from the team member's point of view, *The Team Memory Jogger™* goes beyond basic theories to provide you with practical nuts-and-bolts action steps on preparing to be an effective team member, how to get a good start, get work done in teams, and when and how to end a project. *The Team Memory Jogger™* also teaches you how to deal with problems that can arise within a team. It's perfect for all employees at all levels.

Code: 1050E Price: ***$7.95***

Quantity discounts are available. Please call for details!

Ordering Information: 5 Ways to Order

CALL

TOLL FREE

1-800-643-4316
or 603-893-1944
8:30 AM – 5:00 PM EST

MAIL

GOAL/QPC
2 Manor Parkway
Salem, NH
03079-2841

WEB SITE

www.goalqpc.com

FAX

603-870-9122
Any Day, Any Time

E-MAIL

service@goal.com
Any Day, Any Time

Price Per Copy

Quantity	Price
1–9	$7.95
10–49	$6.95
50–99	$6.25
100–499	$5.75
500–1999	$5.50

For quantities of 2000 or more, call for a quote.

Sales Tax

Canada	7% of order
Georgia	Applicable county tax

Shipping & Handling Charges

Continental US: Orders up to $10 = $2 (US Mail). Orders $10 or more = $4 + 4% of order (guaranteed Ground Delivery). Call for Overnight and Second Day. **For Alaska, Hawaii, Canada, Puerto Rico and other countries, please call.**

Payment Methods

We accept payment by check, money order, credit card, or purchase order. **If you pay by purchase order**: 1) Provide the name and address of the person to be billed, or 2) Send a copy of the P.O. when order is payable by an agency of the federal government.

Order Form for The Memory Jogger™II

1. Shipping Address (We cannot ship to a P.O. Box)

Name ______________________________

Title ______________________________

Company ______________________________

Address ______________________________

City ______________________________

State __________ Zip ____________ Country ___________

Phone ________________ Fax ________________

E-mail ______________________________

2. Quantity & Price

Code	Quantity	Unit Price	Total Price
1030E			
		Tax GA & Canada only	
		Shipping & Handling See opposite page	
		Total	

3. Payment Method

❒ Check enclosed (payable to GOAL/QPC) $ ___________

❒ VISA ❒ MasterCard ❒ Amex ❒ Diners Club ❒ Discover

Card # ____________________ Exp. date __________

Signature ______________________________

❒ Purchase order # ______________________________

Bill to ______________________________

Address ______________________________

City ______________________________

State __________ Zip ____________ Country ___________

4. Request for Other Materials

❒ Information on products, courses & training

❒ Information on customization

000X2

We'd Like to Know What You Think of The Memory Jogger™ II . . .

Your opinions about this product are important to us. Please return your completed survey by mail or fax to GOAL/QPC, attention Memory Jogger II. Thank you!

1. **How did you hear about this book?**

- ❑ GOAL/QPC Product Catalog
- ❑ At a conference/expo
- ❑ Coworker
- ❑ Magazine advertisement
- ❑ Purchased original *Memory Jogger*™
- ❑ Other ____________________

2. **What do you like most about this book?**

3. **How will you use this book? Check all that apply.**

- ❑ Training class text
- ❑ Personal reference
- ❑ Post-training reference
- ❑ Coaching/mentoring
- ❑ Other ____________________

4. **How can we make this a better product for you?**

- ❑ Please add my name to your mailing list.
- ❑ I prefer not to be added to your mailing list.

Name ____________________

Title ____________________

Company ____________________

Address ____________________

City ____________________

State ______ Zip ________ Country ____________

Phone ____________ Fax ____________

E-mail ____________________